Springer Theses

Recognizing Outstanding Ph.D. Research

For further volumes:
http://www.springer.com/series/8790

Aims and Scope

The series "Springer Theses" brings together a selection of the very best Ph.D. theses from around the world and across the physical sciences. Nominated and endorsed by two recognized specialists, each published volume has been selected for its scientific excellence and the high impact of its contents for the pertinent field of research. For greater accessibility to non-specialists, the published versions include an extended introduction, as well as a foreword by the student's supervisor explaining the special relevance of the work for the field. As a whole, the series will provide a valuable resource both for newcomers to the research fields described, and for other scientists seeking detailed background information on special questions. Finally, it provides an accredited documentation of the valuable contributions made by today's younger generation of scientists.

Theses are accepted into the series by invited nomination only and must fulfill all of the following criteria

- They must be written in good English.
- The topic should fall within the confines of Chemistry, Physics, Earth Sciences, Engineering and related interdisciplinary fields such as Materials, Nanoscience, Chemical Engineering, Complex Systems and Biophysics.
- The work reported in the thesis must represent a significant scientific advance.
- If the thesis includes previously published material, permission to reproduce this must be gained from the respective copyright holder.
- They must have been examined and passed during the 12 months prior to nomination.
- Each thesis should include a foreword by the supervisor outlining the significance of its content.
- The theses should have a clearly defined structure including an introduction accessible to scientists not expert in that particular field.

Koutarou Kyutoku

The Black Hole–Neutron Star Binary Merger in Full General Relativity

Dependence on Neutron Star Equations of State

Doctoral Thesis accepted by
Kyoto University, Japan

Springer

Author (Current address)
Dr. Koutarou Kyutoku
Cosmophysics Group
KEK, IPNS, Theory Center
Ibaraki
Japan

Supervisor
Prof. Masaru Shibata
Kyoto University
Kyoto
Japan

ISSN 2190-5053 ISSN 2190-5061 (electronic)
ISBN 978-4-431-54200-1 ISBN 978-4-431-54201-8 (eBook)
DOI 10.1007/978-4-431-54201-8
Springer Tokyo Heidelberg New York Dordrecht London

Library of Congress Control Number: 2012953379

Printed on acid-free paper

Springer is part of Springer Science+Business Media (www.springer.com)

Parts of this thesis have been published in the journal articles:

1. K. Kyutoku, M. Shibata, K. Taniguchi, Phys. Rev. D **82**, 044049 (2010).

2. K. Kyutoku, H. Okawa, M. Shibata, K. Taniguchi, Phys. Rev. D **84**, 064018 (2011), and are reprinted with permission.

Supervisor's Foreword

One of the most exciting events in general relativity and astrophysics in this decade will be the first direct detection of gravitational waves. This will be achieved by 2020. After their first detection, gravitational-wave astronomy will begin and will be a new tool for observing general relativistic objects which have been poorly explored. The black hole–neutron star binaries, which have not been observed yet, are among the most promising sources of gravitational waves. The first detection of them will be achieved by gravitational-wave observation, and we expect that it will provide rich information for black hole–neutron star binaries and properties of the neutron stars.

The merger remnant of black hole–neutron star binaries is also the leading candidate for the central engines of short-duration gamma-ray bursts, for which the progenitor has not been determined yet. A coincident observation of gravitational waves and gamma-ray bursts could provide the definite answer for this unsolved issue. In addition, a class of black hole–neutron star binaries is likely to eject neutron-rich material through the tidal disruption event of the neutron star. The ejected material is a promising transient source of UV, optical, and radio signals, which also have not yet been detected. Observation of these electromagnetic signals is one of the exciting unsolved issues in astronomy, and will be an important method for exploring the black hole–neutron star binaries.

In near-future observations by gravitational-wave detectors and electromagnetic telescopes, black hole–neutron star binaries will be explored in detail. However, to extract physical and astrophysical information from the observational data, a theory for their merger process is necessary. Numerical relativity is probably a unique approach for determining the nature of the merger of black hole–neutron star binaries.

Numerical relativity is the field in which Einstein's equation and matter equations are numerically and accurately solved in computers. Because the merger processes are highly dynamical and general relativistic, numerical relativity is required. Numerical relativity has been significantly developed in particular in the past decade. Now, it is feasible to perform numerical-relativity simulations for a variety of problems such as mergers of binary neutron stars and binary black holes.

Among many others, the simulation for black hole–neutron star binaries was a new topic in this field. The first simulation had been performed in 2006. Since then, significant progress has been achieved in this community and Dr. Koutarou Kyutoku, the author of this volume, has been a central person in this progress.

Dr. Kyutoku is one of the first persons who performed numerical-relativity simulations for the merger of black hole–neutron star binaries systematically. In this problem, there are several free parameters: masses of the black hole and neutron star and black-hole spin. In addition, the equation of state of neutron stars is still poorly known. Thus, the simulation has to be performed for a variety of possible equations of state. Dr. Kyutoku performed a large number of simulations for a variety of parameter sets and equations of state, and clarified the nature of the merger process, merger remnants, and emitted gravitational waves in a comprehensive manner. This volume presents the results of such a systematic investigation. In particular, he pointed out the possibility that the equation of state for neutron stars, which is poorly known, could be strongly constrained by the observation of high-frequency gravitational waves for the first time. He also clarified that for a class of black hole–neutron star binaries, the remnant is composed of a rapidly spinning black hole surrounded by a massive and dense torus. Such a remnant is a promising candidate for the central engine of short-duration gamma-ray bursts. These findings give readers new insights in general relativity and high-energy astrophysics.

Kyoto, September 2012 Masaru Shibata

Acknowledgments

I am deeply grateful to Prof. Masaru Shibata, my supervisor, for his continuous help and support in all the aspects of my research. I am also grateful to him for providing me plentiful computational resources.

I thank Prof. Keisuke Taniguchi for helping me with computing quasiequilibrium states of binaries with LORENE. I also thank Dr. Tetsuro Yamamoto for developing the code SACRA, and Dr. Hirotada Okawa for the update and parallelization of SACRA as well as for organizing our computational environments.

I would like to express my gratitude to Prof. Takashi Nakamura, Prof. Takahiro Tanaka, Prof. Masaki Ando, and Prof. Naoki Seto for discussions of gravitational-wave astronomy on a weekly basis. I am especially indebted to Prof. Naoki Seto for careful reading of the manuscript.

I am grateful to Dr. Yuichiro Sekiguchi, Dr. Kenta Kiuchi, Dr. Yudai Suwa, and Mr. Kenta Hotokezaka for in-depth discussions of numerical relativity, high-energy astrophysics, and physics of neutron stars. I am also grateful to Dr. Atsushi Naruko and Dr. Kentaro Tanabe for daily discussions of a wide range of physics.

It is a pleasure for me to acknowledge Prof. John L. Friedman, Prof. Eric Gourgoulhon, Dr. Nicolas Vasset, and Dr. Benjamin D. Lackey for their helpful discussion. I had an extraordinary time when I visited their institutes.

Kyoto, January 2012 Koutarou Kyutoku

[illegible]

[illegible]

[illegible] the present [illegible]

[illegible]

[illegible]

Contents

Chapter 1
Introduction

The merger of black hole (BH)–neutron star (NS) binaries will give us a unique opportunity to explore many aspects of unknown physics in the near future. Gravitational waves from the merger of such binaries will tell us invaluable information of the NS properties, especially of the equation of state (EOS) at nuclear and supranuclear density. In particular, the EOS strongly modifies the gravitational waveform when the NS is tidally disrupted by the tidal force field of the BH before they merge. A BH–hot, massive accretion disk system is naturally formed after the tidal disruption, and this system will launch a high-energy jet, which may be observed as a short-hard gamma-ray burst (GRB). If cold, neutron-rich material in the NS is ejected from the system after the tidal disruption, the material may accompany r (rapid neutron capture)-process nucleosynthesis. In this chapter, we review physics associated with the BH–NS binary merger, briefly summarize a history of BH–NS studies, and present the purpose of this thesis.

1.1 Gravitational Waves from Compact Binaries

Coalescing binaries composed of a BH and/or a NS, which are frequently called compact binaries, are among the most promising sources of gravitational waves for ground-based laser-interferometric gravitational-wave detectors such as LIGO [1] and Virgo [2]. Gravitational waves are propagation of the spacetime curvature, and are one of the most distinguishing predictions of general relativity. The existence of gravitational waves is confirmed indirectly by observing orbital decay of the Hulse-Taylor binary pulsar, PSR B1913 + 16 [3]. Although we have not observed gravitational waves directly yet (at the end of 2011), detections of gravitational waves will be accomplished in a decade to come by planned next-generation gravitational-wave detectors such as advanced LIGO, advanced Virgo, and KAGRA (formerly LCGT) [4]. Namely, the era of gravitational-wave astronomy will begin [5]. Indeed, current gravitational-wave detectors already begin to present scientifically meaningful results using the absence of gravitational-wave detection [6–9].

K. Kyutoku, *The Black Hole–Neutron Star Binary Merger in Full General Relativity*, Springer Theses, DOI: 10.1007/978-4-431-54201-8_1, © Springer Japan 2013

Because gravitational waves are much more transparent to the absorption and scattering by material than electromagnetic waves and even neutrinos are, gravitational-wave astronomy is expected to become a powerful and unique way to observe strongly gravitating phenomena in our Universe. Among such phenomena, the merger of a BH–NS binary plays an important role to investigate properties of the NS such as the radius and the EOS of a high-density nuclear matter [10–15], and will have a significant impact on astrophysics and nuclear physics. Precise knowledge of the NS EOS is also important to study the EOS of dark energy via sole gravitational-wave observation of a relation between the luminosity distance and the cosmological redshift [16], using a planned third-generation gravitational-wave detector, the Einstein Telescope [17]. An important constraint on the NS EOS is obtained from detection of a $1.97 \pm 0.04 M_\odot$ NS, which is the most massive NS currently known, by a pulsar-timing observation [18]. However, we still do not know the realistic EOS of the NS, because there is no robust measurement of the NS radius (see Sect. 2.3.3). To determine or at least constrain the NS radius and EOS by observing gravitational waves from the BH–NS binary, we have to prepare accurate theoretical templates of gravitational waveforms employing a wide variety of the NS EOSs and other physical parameters.

The coalescence of compact binaries is usually classified into three phases, i.e., the inspiral phase, merger phase, and ringdown phase. Compact binaries are believed to be formed mainly via two supernova explosions in a binary system [19], and formed binaries gradually contract due to the gravitational radiation reaction. The inspiral phase denotes the phase in which the radiation reaction timescale is much longer than the orbital period (see below), so that the adiabatic approximation holds well. In the inspiral phase, orbital evolution and emitted gravitational waves are computed accurately by a post-Newtonian (PN) approximation (see [20] and references therein for reviews). It is also important that a point-particle approximation also holds well in the early inspiral phase, because the orbital separation is much larger than the size of each object. As the binary separation decreases, the radiation reaction time scale becomes comparable or even shorter than the orbital period, and also the finite size effect becomes important. This phase is called the merger phase, and a computer simulation by numerical relativity is the unique approach to investigate the merger phase, because the nonlinearity of strong gravity and hydrodynamics play important roles. The remnant BH left after the merger[1] emits gravitational waves associated with the quasinormal-mode oscillation, and settles into a stationary BH. This phase is called the ringdown phase, and the frequency and damping time scale of the quasinormal mode are computed by the BH perturbation technique (see [24] and reference therein for reviews).

[1] For the merger of binary NSs, a frequent outcome will not be a BH, but be a hypermassive NS [21–23]. If the hypermassive NS is formed, gravitational waves just after the merger are not emitted by the BH ringdown, but by oscillation of the hypermassive NS. We do not go into detail of the hypermassive NS, because it is not relevant to the BH–NS binary. It should be noted that gravitational waves from hypermassive NSs will tell us information at higher density than at the central density of a canonical NS.

1.1.1 *The Quadrupole Formula and the Inspiral Phase*

It is worthwhile to review basic properties of gravitational waves derived by the lowest-order quadrupole formula [25], which is sufficient to describe semiquantitative properties of the compact binary in the inspiral phase. Here, we consider a circular binary consists of a BH with the mass $M_{\rm BH}$ and a NS with the mass $M_{\rm NS}$, and denote the mass ratio and the total mass by $Q \equiv M_{\rm BH}/M_{\rm NS}$ and $m_0 \equiv M_{\rm BH}+M_{\rm NS}$, respectively. These symbols are used throughout this thesis. We denote the gravitational constant and the speed of light by G and c, respectively[2] throughout this thesis. The orbital angular velocity for a given binary separation d is

$$\Omega = \sqrt{\frac{Gm_0}{d^3}}, \tag{1.1}$$

and hence the orbital period is given by

$$P_{\rm orb} \equiv \frac{2\pi}{\Omega} = 2\pi\sqrt{\frac{d^3}{Gm_0}}, \tag{1.2}$$

in Newtonian gravity. The quadrupole formula gives gravitational waves in the transverse-traceless gauge in terms of a mass quadrupole moment as

$$h_{ij}^{\rm TT} = \frac{2G}{c^4 r}\left[\frac{{\rm d}^2}{{\rm d}t^2}\mathcal{Q}_{ij}^{\rm TT}\left(t-\frac{r}{c}\right)\right], \tag{1.3}$$

$$\mathcal{Q}_{ij} \equiv \int \rho\left(x_i x_j - \frac{1}{3}r^2 f_{ij}\right){\rm d}^3x, \tag{1.4}$$

where ρ, r, and f_{ij} denote the rest-mass density, the coordinate distance from the center of mass of the binary $|x^i|$, and the flat three metric, respectively. In this section, we assume that lowercase Latin indices denote Cartesian components, and therefore we do not distinguish contravariant and covariant indices. We obey the Einstein summation convention throughout this thesis, i.e., we always take the sum of repeated indices for a pair of contravariant and covariant indices, and we also take the sum of repeated covariant indices in this section. The superscript "TT" means that the quantity is evaluated by a transverse-traceless projection using a unit vector $\hat{x}_i \equiv x_i/r$ as

$$\mathcal{Q}_{ij}^{\rm TT} = \left(P_{ik}P_{jl} - \frac{1}{2}P_{ij}P_{kl}\right)\mathcal{Q}_{kl}, \tag{1.5}$$

$$P_{ij} = f_{ij} - \hat{x}_i\hat{x}_j. \tag{1.6}$$

---

[2] $G = 6.67\times10^{-8}\,{\rm g^{-1}\,cm^3\,s^{-2}}$, $c = 3.00\times10^{10}\,{\rm cm\,s^{-1}}$.

In the quadrupole approximation, the gravitational-wave luminosity is given by

$$\left(\frac{\mathrm{d}E}{\mathrm{d}t}\right)_{\mathrm{GW}} = \frac{G}{5c^4}\left\langle \frac{\mathrm{d}^3 \mathcal{Q}_{ij}^{\mathrm{TT}}}{\mathrm{d}t^3}\frac{\mathrm{d}^3 \mathcal{Q}_{ij}^{\mathrm{TT}}}{\mathrm{d}t^3}\right\rangle, \tag{1.7}$$

where the bracket stands for a value averaged over several wavelengths, and the gravitational-wave luminosity from a circular binary is derived as

$$\left(\frac{\mathrm{d}E}{\mathrm{d}t}\right)_{\mathrm{GW}} = \frac{32G^4}{5c^5}\frac{m_0 M_{\mathrm{BH}}^2 M_{\mathrm{NS}}^2}{d^5}. \tag{1.8}$$

Using the fact that the orbital binding energy of a circular binary is given by

$$E = -\frac{GM_{\mathrm{BH}}M_{\mathrm{NS}}}{2d}, \tag{1.9}$$

the lifetime of the binary is formally defined as the time required for the orbital separation, d, becomes zero by

$$t_{\mathrm{GW}} = \frac{5c^5}{256G^3}\frac{d^4}{M_{\mathrm{BH}}M_{\mathrm{NS}}m_0} \tag{1.10}$$

$$\simeq 14\left(\frac{d}{9\times 10^6\,\mathrm{km}}\right)^4\left(\frac{Q(1+Q)}{3\times 4}\right)^{-1}\left(\frac{M_{\mathrm{NS}}}{1.35M_{\odot}}\right)^{-3}\ \mathrm{Gyr}, \tag{1.11}$$

assuming that all the formulae are valid up to $d = 0$ [26, 27]. Therefore, a compact binary with $d \lesssim 10^7$ km merges within the age of the universe. Whereas a BH–NS binary and a stellar-mass binary BHs have never been observed yet, many of the already known binary NSs satisfy this condition [28].

For an elliptic orbit with a semimajor diameter d and an orbital eccentricity e, the emission rate of energy and angular momentum is enhanced as

$$\left(\frac{\mathrm{d}E}{\mathrm{d}t}\right)_{\mathrm{GW}} = \left(\frac{\mathrm{d}E}{\mathrm{d}t}\right)_{\mathrm{GW},e=0}\left(1+\frac{73}{24}e^2+\frac{37}{96}e^4\right)(1-e^2)^{-7/2}, \tag{1.12}$$

$$\left(\frac{\mathrm{d}J}{\mathrm{d}t}\right)_{\mathrm{GW}} = \left(\frac{\mathrm{d}J}{\mathrm{d}t}\right)_{\mathrm{GW},e=0}\left(1+\frac{7}{8}e^2\right)(1-e^2)^{-2}, \tag{1.13}$$

and the enhancement of the energy loss is more significant than of the angular momentum loss. Recall that the orbital binding energy of a binary is determined only by the masses of components and the semimajor diameter, and the maximum value of the orbital angular momentum is obtained for a circular binary. Using the fact that the eccentricity is related to the orbital binding energy, E, and angular momentum, J, by

$$e^2 = 1 + \frac{2m_0 E J^2}{G^2 M_{\rm BH}^3 M_{\rm NS}^3}. \tag{1.14}$$

it is shown that the orbital eccentricity approaches zero more rapidly than the semi-major diameter does as

$$a(e) = \frac{c_0 e^{12/19}}{1-e^2}\left(1 + \frac{121}{304}e^2\right)^{870/2299} \tag{1.15}$$

$$\simeq c_0 e^{12/19} \ (\text{for } e^2 \ll 1), \tag{1.16}$$

where c_0 is a constant determined by an initial condition. In other words, the gravitational radiation circularizes an elliptic binary orbit, and therefore the binary just before the merger with $d \lesssim 100\,\text{km}$ is safely believed to be circular [26, 27].

1.1.2 The Estimated Detection Rate

Because the BH–NS binary has never been observed yet, the merger rate of the BH–NS binary and the detection rate for ground-based detectors are fairly uncertain. Theoretical estimations of the merger rate, e.g., the merger rate per 100 Myr^{-1} per Milky Way Equivalent Galaxy, relies on the population synthesis (see [19, 29, 30] and references therein for reviews). In the population synthesis, the merger rate is estimated by Monte-Carlo simulations adopting models of stellar evolution, which are calibrated to observations, such as supernova rates, pulsar distributions, and binary NSs. The deficit of this method is a large number of uncertainties in the modeling. For example, the initial mass function of binaries is more uncertain than the initial mass function of single stars is. Main uncertainties of stellar and binary evolution come from the mass-loss rate due to the stellar wind [31] and treatment of the evolution during the common-envelope phase [32]. Another difficulty is conversion from the merger rate to the detection rate. Aside from precise evaluation of the horizon distance (see below) for a given binary configuration, it is pointed out that the elliptic galaxy contributes to the detection rate in a different manner from the spiral galaxy does, because the star formation history is different [33]. As a result of these and other issues, the estimated detection rate can differ by orders of magnitude.

At the end of 2011, the most plausible estimation of the detection rate may be the one shown in [30], in which the authors summarizes an estimation of the compact binary coalescence rate for Initial and Advanced LIGO–Virgo network. In particular, the pessimistic, realistic, and optimistic detection rate of a BH–NS binary coalescence is estimated to be 7×10^{-5}, 0.004, and 0.1 per year for the Initial configuration, respectively, and 0.2, 10, and 300 per year for the Advanced configuration, respectively. Although these values are still highly uncertain, it is sufficient to lead us to believe that the BH–NS binary will be an important target of gravitational-wave astronomy with the Advanced detectors.

Before closing this section, it will be important to recall an expected strength of gravitational waves in the quadrupole approximation. Using (1.3), it is shown that an observed gravitational-wave amplitude at a distance D along the rotational axis from a circular BH–NS binary with the orbital separation d is

$$h = \frac{4GM_{\rm BH}M_{\rm NS}}{c^4 Dd} \tag{1.17}$$

$$\simeq 3.2 \times 10^{-22} \left(\frac{M_{\rm NS}}{1.35M_\odot}\right)\left(\frac{Q/(1+Q)}{3/4}\right)\left(\frac{6m_0}{c^2 d}\right)\left(\frac{D}{100\,{\rm Mpc}}\right)^{-1}, \tag{1.18}$$

and the quadrupole-mode gravitational-wave frequency is given by twice the orbital frequency as

$$f = \frac{\Omega}{\pi} \tag{1.19}$$

$$\simeq 810 \left(\frac{6Gm_0}{c^2 d}\right)^{3/2}\left(\frac{M_{\rm NS}}{1.35M_\odot}\right)^{-1}\left(\frac{1+Q}{4}\right)^{-1} {\rm Hz}. \tag{1.20}$$

The spectral amplitude of gravitational waves in the quadrupole approximation is derived by the stationary phase approximation as

$$|\tilde{h}(f)| = \sqrt{\frac{5}{24}}\frac{G^{5/6}}{\pi^{2/3}c^{3/2}D}\frac{M_{\rm BH}^{1/2}M_{\rm NS}^{1/2}}{m_0^{1/6}} f^{-7/6} \tag{1.21}$$

$$\simeq 4.4 \times 10^{-25} \times \left(\frac{M_{\rm NS}}{1.35M_\odot}\right)^{5/6}\left(\frac{Q^{1/2}/(1+Q)^{1/6}}{3^{1/2}/4^{1/6}}\right)\left(\frac{f}{1\,{\rm kHz}}\right)^{-7/6} {\rm Hz}^{-1}. \tag{1.22}$$

In the gravitational-wave data analysis, detectability of a signal is first estimated by a signal-to-noise ratio (SNR) defined as

$$\rho_{\rm SNR}^2 \equiv 4\int_0^\infty \frac{|\tilde{h}(f)|^2}{S_n(f)} {\rm d}f, \tag{1.23}$$

where $S_n(f)$ denotes the one-sided noise power spectral density of a detector. For a given source, $\rho_{\rm SNR}$ depends on the distance to the source and on the noise power spectral density. When a threshold SNR, which is typically taken to be 8, is specified, the horizon distance for a detector is defined to be the distance at which $\rho_{\rm SNR}$ becomes equal to the threshold SNR. Whereas it is not easy to compute the SNR due to detector-dependent $S_n(f)$, the horizon distance for the BH–NS binary is estimated in [30] to be 70 and 927 Mpc for the Initial and Advanced configurations, respectively, assuming a $10M_\odot$ BH and a $1.4M_\odot$ NS and using $S_n(f)$ obtained in [1]. Figure 1.1 shows the noise amplitude spectral density $\sqrt{S_n(f)}$, which has a dimension of $1/\sqrt{\rm Hz}$, adopted in [30].

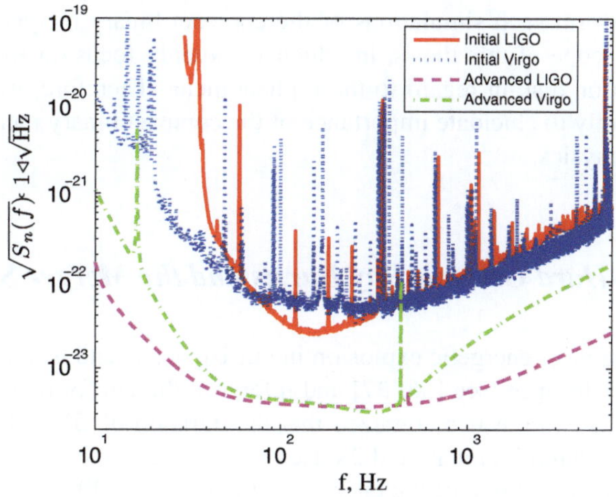

Fig. 1.1 The square root of noise power spectral densities (or the noise amplitude spectral densities) of the Initial LIGO, the Initial Virgo, the Advanced LIGO, and the Advanced Virgo. The power spectral density of the Initial LIGO is obtained by the LIGO S5 run [1], and that of the Initial Virgo is obtained by the Virgo VSR2 run [2]. This figure is taken from [30]

1.2 Electromagnetic Emission from the Compact Binary Merger

The merger of a BH–NS binary is a potential candidate for the progenitor of short-hard GRBs in the so-called merger scenario [34], as well as the binary NSs are. The investigation of the central engine and emission mechanism of the GRB is one of the most important problems in astrophysics. If a short-hard GRB is detected concurrently with gravitational waves from a compact binary, the merger scenario is shown to be highly convincing. Moreover, if electromagnetic counterparts of the compact binary merger is detected, it helps the estimation of binary parameters by unambiguous (or less ambiguous than solely by gravitational waves) determination of the location of the gravitational-wave source on the celestial sphere [35]. The GRB is of course one of the promising counterparts, because the delay time from the merger to the emission is assumed to be negligible. The detection of X-ray, optical, and radio afterglow also helps the location determination, particularly when the GRB is not detected due to the misalignment of the jet axis. Another interesting counterpart is a "kilonova," in which the electromagnetic radiation is emitted isotropically by decay of heavy nuclei synthesized in the *r*-process. A possibility of the *r*-process is itself important to explain the existence of heavy, neutron-rich elements, i.e., *r*-process elements, with $A \gtrsim 90$–110 such as ^{129}I and ^{235}U.

To fully investigate these high-energy phenomena associated with the compact binary merger, it is necessary to perform numerical-relativity simulations incorporating neutrino transports and/or electromagnetic effects. Whereas it is now becoming possible to perform such simulations of the compact binary merger [22, 23], this is beyond the scope of this thesis, in which we mainly focus on formation of the remnant accretion disk in regard to these phenomena. Therefore, we review these issues only briefly to elucidate importance of the compact binary merger for a wide range of astrophysics.

1.2.1 Short-Hard Gamma-Ray Bursts and the Merger Scenario

The GRB is the most energetic explosion in our Universe, and is known to occur at a cosmological distance (see [36, 37] and reference therein for reviews). From the era of *BATSE* satellite, it is recognized that the duration of GRBs have a bimodal distribution with a minimum around 2 s. Here, the duration means the so-called T_{90}, in which intermediate 90 % of the energy is emitted in 20–2000 keV. According to this bimodality, GRBs are usually[3] classified into two categories. One is the long GRB, for which $T_{90} > 2$ s, and the other is the short GRB, for which $T_{90} < 2$ s. Whereas ratios between two classes according to this classification are different for different satellites, such as *Swift*, due to the different performance of instruments, the bimodal character does not change. Here, care must be taken for two facts. First, this classification is not very physical, and penetration from one class in a physical sense into the other class in an observational sense is always unavoidable. Second, T_{90} is determined at the satellite location, and do not represent the intrinsic time duration of the GRB. Another important observational finding is that the spectrum of the short GRB is harder than that of the long GRB by comparing the hardness ratio, which is defined as the ratio of the fluence in 50–100 keV to that in 25–50 keV. Therefore the short GRB is also called the short-hard GRB. These systematic differences lead many researchers to believe that the long and short GRBs represent different classes of astrophysical events.

The mechanism of the GRB is long under debate, and it is widely accepted that the highly relativistic motion of the emission source is necessary to solve the "compactness problem" [38, 39]. First of all, a typical size of the emission source is roughly estimated to be $c\delta t$, where $\delta t \sim$ 10–100 ms is a shortest time scale of variability in the GRB light curve. Next, when the energy flux and distance to the GRB are observed, we can estimate the isotropic-equivalent luminosity, which is typically 10^{50}–10^{52} erg s^{-1} for both long and short GRBs. It should be noted that such large values heavily relies on the fact that the GRB occurs at a cosmological distance.[4] The isotropic-equivalent energy is also estimated by multiplying the burst duration, and the isotropic-equivalent energy is lower for short-hard GRBs as 10^{49}–10^{51} erg than for long GRBs due to a shorter time duration. Finally, using a typical energy of the observed photon, we can estimate a photon number density in the emitting region

---

[3] The existence of intermediate GRBs is also suggested by several observations, but is not conclusive.

[4] We neglect the effect of the cosmological redshift for simplicity. This simplification does not change the conclusion, particularly for short-hard GRBs.

of the GRB. Assuming that the main source of opacity is the Thomson scattering by electron–positron (e^-e^+) pairs,[5] typical optical depth is estimated to be $\sim 10^{13}$, and hence a thermal spectrum is expected. However, the observed spectrum of the GRB is usually nonthermal, and indeed this compactness problem had been thought to be evidence for galactic origins of GRBs before the *BATSE* observation.

The solution to this contradiction is to consider that the GRB source is in a highly relativistic motion. For a high Lorentz factor of $w \gg 1$, the size of the source is larger by a factor of w^2 due to relativistic corrections than naively estimated. A radial correction due to the motion with a velocity $v \approx c(1 - 1/2w^2)$ to us and an angular correction due to the relativistic beaming within an angle $\theta \approx 1/w$ for two points on a spherical shell both modify the size of the emission region to be $\sim w^2 c\delta t$. Another important correction is a Doppler shift of observed photons. Because the photon energy at the source is reduced by a factor of w, the number of photons which can create e^-e^+ pair plasma is also reduced. In particular, if the observed photon has a number distribution $N(E)$ of the form $N(E)dE \propto E^{-\alpha}dE$, the number of photon which can contribute to the pair production is reduced by a factor of $w^{2(\alpha-1)}$ due to the necessity of two photons and integration of distribution functions. Putting it all together, the constraint on the Lorentz factor is obtained by the condition in which the optical depth has to be smaller than unity, and it is usually said that $w \gtrsim 30$ is required to explain the short-hard GRB observation. Notice that this value is smaller than the value assumed for a long GRB, $w \gtrsim 100$. At any rate, central engines of GRBs are required to be able to launch a jet with such a high Lorentz factor.

The long GRBs are now thought to be associated with a death of massive stars [40], because a core-collapse supernova explosion sometimes follows the long GRB. In this model, which is called the collapsar model, the central engine of the GRB is assumed to be a BH–accretion disk system formed as a result of the stellar core collapse, and this model is now thought to be convincing after many theoretical efforts. It is difficult to explain the short-hard GRB by the collapsar model, however, because some of short-hard GRBs are found to be associated with early-type galaxies with low star-formation rates [41, 42]. This problem is not a problem for the merger scenario, in which the inspiral and merger of a binary spend a long time after the death of each component. The merger scenario also naturally explains observation in that the short-hard GRB is found at a closer distance to the earth than the long GRB is, say the cosmological redshift smaller than unity, even when it is not associated with an early-type galaxy. Taking the fact that many properties of the long and short GRBs are still common into account, such as a huge energy and a rapid time variability, it is expected that a similar model of the central engine to the collapsar model explains the short-hard GRB.

The short-hard GRBs (see [39, 43] and references therein for reviews) are thought to originate from the merger of compact binary including NSs, such as a BH–NS binary. The central engine is assumed to be a BH–accretion disk system in the merger scenario, as well as in the collapsar model. According to the purpose of this thesis,

---

[5] For a short-hard GRB, it is suggested that the opacity of e^-e^+ pair production from two photons itself gives a weaker constraint compared to the case of a long GRB [39].

we only focus on the BH–NS binary. If a NS is tidally disrupted during the merger of a BH–NS binary, a system composed of a spinning BH and a hot, massive accretion disk of $\gtrsim 0.01 M_\odot$ may be formed. Such an outcome could be a central engine of the GRB, because it could radiate a large amount of energy $\gtrsim 10^{48}$ erg in a short time scale of $\lesssim 2$ s to launch a GRB jet if the accretion rate is high. This amount of energy may be sufficient to explain the short GRB with a beaming effect, whereas it have to be noted that the opening angle of the short-hard GRB jet is more uncertain than that of the long GRB jet. One plausible mechanism of the GRB jet formation is neutrino–anti neutrino ($\nu\bar{\nu}$) annihilation and subsequent e^-e^+ pair production process [34, 44, 45]. If both temperature and density are high in the accretion disk, the neutrino cooling via e^+/e^- capture will dominate over radiation and advection cooling [46]. The neutrino luminosity can become large if the accretion rate is as high as $\gg 0.1 M_\odot\ \mathrm{s}^{-1}$, and total efficiency of converting accreting masses first to the neutrino emission and next to the e^-e^+ pair plasma is estimated to be $\sim 10^{-4}$ in optimal cases by several calculations [39]. Therefore, the required mass of the accretion disk is $\gtrsim 0.01 M_\odot$. Another mechanism is the magnetohydrodynamic process such as the Blandford-Znajek mechanism [47, 48]. In this model, magnetized plasma is ejected from the accretion disk into the funnel region around the rotational axis, and a Poynting-flux dominated jet is launched along the rotational axis. This mechanism is shown to have a higher efficiency depending on the BH spin than the neutrino mechanism [49], and therefore it will be easier to explain the GRB if sufficient (in strength and geometry) magnetic fields are provided by the accretion disk to drive the MHD jet.

Here, both mechanisms clearly require formation of an accretion disk as a result of tidal disruption to drive a GRB jet. Only numerical relativity can answer quantitatively the question whether/when the formation of a massive accretion disk is possible for the BH–NS binary merger.

1.2.2 The r-Process and the Kilonova

The possibility of the r-process in the BH–NS binary merger has long been investigated [50]. The r-process is a neutron capture process faster than the β-decay, and is thought to be responsible for producing about a half of heavy, neutron-rich nuclei above iron along the neutron-drip line (see [51, 52] and references therein for reviews). Because a fairly high neutron flux is required for the neutron capture to proceed faster than the β-decay, an astrophysical site where the r-process becomes possible is a matter of debate for a long time. In particular, the solar abundance shows a double-peaked structure around $A \approx 130$ and 195, and therefore theoretical models of the r-process have to reproduce this structure. It has been proposed that the neutrino-driven wind from a newly-born NS after the supernova will produce a neutron-rich, high-entropy material so that the r-process occur [53], but this model is found to reproduce only the first peak at $A \approx 130$ and do not reproduce the structure above it, particularly the second peak at $A \approx 195$, in typical situations. Generally speaking, the requirements for the r-process site are high entropy, a low electron fraction, and a short dynamic time scale. If the NS material is ejected from the BH–NS binary and becomes unbound as a result of tidal disruption, a very low electron fraction of the NS material may provide a promising site of the *r*-process.

Radioactive decay of the *r*-process nuclei, if it is formed, is naturally observed as an optical–UV transient event about a day after the merger, depending on the mass of ejected material [54]. This event is named a "kilonova" in [55], because it is brighter by a factor of $\sim 10^3$ than a nova. An advantage of the kilonova is the fact that isotropic emission is expected. Because the GRB and afterglow associated to it will not always be observed for the BH–NS binary merger due to the misalignment of the jet axis, the kilonova may be an important counterpart for the localization of the gravitational-wave source [35]. Again, only numerical relativity is a unique tool to investigate whether the ejected material becomes unbound and whether the physical condition enables the ejected material to be a *r*-process site.

1.3 The Mass-Shedding Limit

The final fate of the BH–NS binary is classified into two categories.[6] One is the case in which the NS is not disrupted before the merger, and the NS is only swallowed by the BH to form a massive, rapidly-spinning remnant BH. In this case, the NS EOS affects gravitational waves only very weakly, and the GRB jet will not be launched due to the lack of the energy budget. The other is the case in which the NS is disrupted before the merger, and some of the disrupted material forms a hot, massive accretion disk around the remnant BH. In this case, the NS EOS strongly modifies gravitational waves in the late inspiral and merger phases, and the GRB jet is expected if the mass of the accretion disk is sufficient. A large portion of the NS is swallowed by the BH even if tidal disruption occurs, and the remnant BH becomes massive and rapidly-spinning.

According to the above consideration, it is interesting to know an approximate condition for tidal disruption to occur. Because it is difficult to define quantitatively the tidal disruption, the mass shedding from the NS should be investigated as a necessary condition of the tidal disruption. In context of the BH–NS binary merger, the mass-shedding limit is defined as the point at which the self-gravity of the NS and the BH tidal force become equal on the NS surface closest to the BH. Here for simplicity, we neglect spins of each object and assume Newtonian gravity. Taking tidal deformation of the NS into account, the mass-shedding limit is given in terms of a binary separation $d_{\rm shed}$ by

[6] The possibility of a stable mass transfer is also suggested [56]. The stable mass transfer occurs if the orbital separation increases faster than the NS radius increases when the mass is tidally stripped from the NS (lighter component) to the BH (heavier component). We do not discuss the stable mass transfer in more detail, because it has never been observed in numerical-relativity simulations. The reason for this may be that the stable mass transfer requires a large mass ratio, whereas the mass shedding requires a small mass ratio especially in full general relativity [57]. It does not exclude the stable mass transfer for a BH–NS binary with a massive and rapidly-spinning BH.

$$\frac{2GM_{\rm BH}(c_{\rm R}R_{\rm NS})}{d_{\rm shed}^3} = \frac{GM_{\rm NS}}{(c_{\rm R}R_{\rm NS})^2}, \tag{1.24}$$

where $R_{\rm NS}$ and $c_{\rm R}$ are the radius of the NS in the absence of tidal effect and a factor of order unity denoting the tidal deformation, respectively. For a while, we only focus on the dependence of the mass-shedding limit on physical quantities to derive qualitative property of the mass-shedding limit. It is shown that the mass-shedding limit in terms of a binary separation is written by[7]

$$d_{\rm shed} \propto Q^{-2/3}\mathscr{C}^{-1}M_{\rm BH}, \tag{1.25}$$

where $\mathscr{C} \equiv GM_{\rm NS}/(c^2R_{\rm NS})$ is the compactness of the NS. It is also useful to rewrite this relation in terms of a normalized orbital angular velocity as

$$\frac{G\Omega m_0}{c^3} \propto \frac{\mathscr{C}^{3/2}(1+Q)^{3/2}}{\sqrt{Q}}, \tag{1.26}$$

assuming a circular orbit.

The mass-shedding limit should be compared to the innermost stable circular orbit (ISCO) of the BH, inside which no material have a stable circular orbit around the BH. The ISCO radius[8] of the BH on the equatorial plane $R_{\rm ISCO}$ is proportional to its mass, $M_{\rm BH}$, and the prefactor depends on the spin angular momentum of the BH $S_{\rm BH}$ [58]. For example, the ISCO radius for a Schwarzschild BH is given by $R_{\rm ISCO} = 6GM_{\rm BH}/c^2$, and that for a extremely spinning Kerr BH is given by $R_{\rm BH} = GM_{\rm BH}/c^2$ for a prograde orbit. Figure 1.2 shows the ISCO radius as a function of the nondimensional spin parameter $a \equiv cS_{\rm BH}/(GM_{\rm BH}^2)$, as well as the radius of the event horizon, which is given by

$$r_+ = \frac{GM_{\rm BH}}{c^2}\left(1+\sqrt{1-a^2}\right). \tag{1.27}$$

By writing the ISCO radius using a spin-dependent parameter ζ as $R_{\rm ISCO} = \zeta GM_{\rm BH}/c^2$, (1.25) is rewritten by

$$\frac{d_{\rm shed}}{R_{\rm ISCO}} \propto Q^{-2/3}\mathscr{C}^{-1}\zeta^{-1}. \tag{1.28}$$

If the value on the left-hand side is larger than unity, the mass shedding occurs outside the BH ISCO, and therefore the accretion-disk formation is expected. Aside from

[7] We essentially compare the mass density of each object.

[8] In this thesis, "the ISCO radius" always represents "the ISCO radius in the Boyer-Lindquist coordinates," which is physical in the sense that it gives the proper circumferential length for the equatorial circular orbit. It should be noted that the coordinate radius of the ISCO in numerical-relativity simulation is different from the Boyer-Lindquist one.

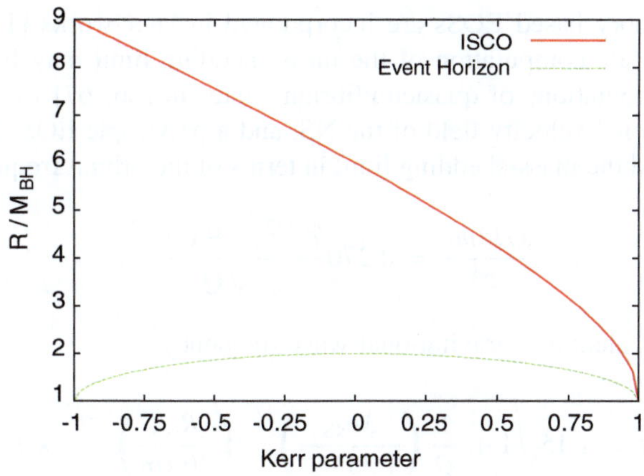

Fig. 1.2 The radius of the ISCO and the event horizon of the BH. The vertical axis denotes the radius normalized by the BH mass, $c^2R/(GM_{\rm BH})$, and the horizontal axis denotes the nondimensional spin parameter of the BH, $cS_{\rm BH}/(GM_{\rm BH}^2)$

a factor of order unity difference, this expression indicates that the mass shedding outside the ISCO becomes easier for the BH–NS binary when

1. The mass ratio of the BH mass to the NS mass is small. Namely, the BH mass should be small.
2. The compactness of NS is small, and hence the NS radius should be large.
3. The spin angular momentum of the BH is large in the magnitude and parallel with the orbital angular momentum. Notice that ζ is a decreasing function of a for a prograde orbit.

The first and second condition states that the tidal effect is the finite size effect of the NS, and the effect becomes strong when the size of the NS ($= R_{\rm NS}$) is larger compared to the size of the BH ($\propto M_{\rm BH}$) for a fixed value of the NS mass.

Quantitative derivation of the mass-shedding limit requires hydrodynamic computations in curved spacetimes. Earlier studies assume the geodesic motion of the NS center in the Kerr spacetime, and solve the Newtonian hydrostatics for an orbiting NS with the tidal force tensor of the Kerr spacetime as an external force [59–62]. In these works, NS EOSs are chosen to be incompressible and the radiation reaction are neglected. Therefore, care must be taken for the fact that the approximation become inaccurate when the BH mass is comparable to the NS mass. The BH mass which can cause the mass shedding is derived as

$$M_{\rm BH} = 4.7 \left(\frac{M_{\rm NS}}{1.35M_\odot}\right)^{-1/2} \left(\frac{R_{\rm NS}}{10\,{\rm km}}\right)^{3/2} \left(\frac{\zeta}{6}\right)^{-3/2} \tag{1.29}$$

in [62]. Similar results are obtained in [63] for polytropic EOSs, and it is found that the mass-shedding limit depends on the EOS. The effect of higher-order gravity and nuclear-theory based EOSs are incorporated in later works [13–15, 64, 65]. The most accurate computation of the mass-shedding limit may be fully general relativistic computations of quasiequilibrium states in [66, 67] for a nonspinning BH, an irrotational velocity field of the NS, and a polytropic EOS. The authors of [66, 67] derived the mass-shedding limit in terms of the orbital frequency as

$$\frac{G\Omega m_0}{c^3} = 0.270\frac{\mathscr{C}^{3/2}(1+Q)^{3/2}}{\sqrt{Q}}, \tag{1.30}$$

and this is equivalent to a gravitational-wave frequency

$$f = 1.15\sqrt{1+\frac{1}{Q}}\left(\frac{M_{\rm NS}}{1.35M_\odot}\right)^{1/2}\left(\frac{R_{\rm NS}}{10\,{\rm km}}\right)^{-3/2} \ {\rm kHz}. \tag{1.31}$$

Notice that $\sqrt{1+1/Q}$ never falls below unity. Taking the fact that the mass shedding outside the BH ISCO is rare for a massive BH, this relation suggests that the gravitational-wave frequency at the mass shedding is typically larger than 1 kHz, above which the sensitivity of ground-based detector becomes worse. The frequency at the tidal disruption is always higher than that at the mass shedding [68], and hence the sensitivity is much worse. To observe gravitational waves at the tidal disruption, it is highly preferable to improve sensitivity at kHz range.

1.4 Black Hole–Neutron Star Binaries in Numerical Relativity

General relativistic computations of quasiequilibrium states of the BH–NS binary are initiated from an approximate treatment of the BH and a corotational velocity field of the NS. An early, unpublished work of Miller [69] adopts the formalism similar to the puncture method [70], whereas the treatment of the singularity is approximate and the shift vector is determined by the minimal distortion gauge condition [71], not by a modern quasiequilibrium condition. A work of Baumgarte et al. [72] solves equations only in a neighborhood of the NS on a background Schwarzschild spacetime written in Kerr-Schild coordinates, assuming that the BH mass is much larger than the NS mass. A corotational velocity field of the NS is assumed in Baumgarte et al. [72], and this method is extended to the irrotational velocity field of the NS and the isotropic coordinate of the Schwarzschild BH in Taniguchi et al. [73]. Fully general relativistic computations are done by several authors adopting the excision method to handle the BH in a self-consistent manner, using a spectral method for the numerical accuracy (see [74] and references therein for reviews). In a work of Grandclément [75, 76], a flat background for the conformal metric (see Sect. 3.2) is adopted, and the boundary condition on the horizon is slightly artificial [77, 78]. A work done by Taniguchi et al. [79] adopts the nonrotating Kerr-Schild background for the conformal metric, and the flat conformal metric is chosen by the same authors later [66]. The boundary condition on the horizon in the works of Taniguchi et al. [66, 79] are essentially derived in the isolated horizon framework [80, 81], but the rotational state of the BH is not controlled and corresponds to the corotational one. The irrotational condition for the BH is imposed in the work of Grandclément [75, 76], and is also imposed in the latest work of Taniguchi et al. [67]. A work of Foucart et al. [82] is done by similar methods to those of Taniguchi et al. [67], and adopt a superposed Kerr-Schild conformal metric [83] as well as the flat one. The authors also implement an eccentricity reduction procedure [84] for the BH–NS binary. A quasiequilibrium sequence of the BH–NS binary is also computed in the puncture framework by the author of this thesis [85], improving the method developed for a corotational velocity field in works of Shibata and Uryū [86, 87] (see also [68]) and adopting new condition to determine the location of the rotational axis (see Sect. 3.4.3). All of these works adopt the polytropic EOS with $\Gamma = 2$, and the effect of the BH spin is not taken into account. These issues are improved in the computation of initial data for numerical simulations.

Newtonian and pseudo-Newtonian simulations of the BH–NS binary mergers are performed in several works. These simulations mainly focus on aspects other than gravitational waves. For example, purely Newtonian (without radiation reaction) smoothed-particle hydrodynamics (SPH) simulations with a polytropic EOS are performed in works of Kluźniak and Lee (or Lee and Kluźniak) [88, 89] and subsequent works by the same authors to investigate dynamical mass transfer in binaries with corotational and irrotational NSs. In a work of Janka et al. [90], the Lattimer-Swesty EOS [91] is adopted for a mesh-based numerical hydrodynamics, radiation reaction terms are added, and neutrino emission is investigated incorporating microphysics for corotational, irrotational, and counter-rotating NSs. Dynamics of accretion disks are also investigated by Newtonian SPH simulations in a work of Rosswog et al. [92] for corotational NSs, incorporating microphysics and radiation reaction. Later, this work is improved to include a stronger effect of gravity using a pseudo-Newtonian potential and to compute irrotational NSs by Rosswog [93]. A similar prescription of gravity to that of [93] is also adopted in a work of Ruffert and Janka [94] with a mesh-based numerical hydrodynamics with an ideal-gas EOS, and a nearly-extremal BH spin is taken into account. Some works partially incorporate effects of full general relativity in an approximate manner. In works done by Faber et al. [95, 96], SPH simulations are performed in the conformal flatness approximation for the gravity [97, 98] with a cold, polytropic EOS, and both corotational and irrotational velocity NSs are adopted for initial conditions. Another SPH simulation is performed in a work of Rantsiou et al. [99] on the background Kerr spacetime in the Kerr-Schild coordinates, and the effect of the BH spin is investigated. In this work, a polytropic EOS is adopted and the corotational NS is assumed for the initial condition. Other approximate evolution scheme is proposed in a work of Soupuerta et al. [100], but this work does not track orbital evolution up to the merger. Generally speaking, the mass of the accretion disk becomes smaller as the treatment of gravity becomes more accurate due to a stronger effect of gravitational attraction.

Fully general relativistic simulations of the BH–NS binary merger have achieved progress in recent years. The first merger simulation is performed in the works of Shibata and Uryū [86, 87], where the NS fluid field is assumed to be corotational in the initial condition. This is updated to the irrotational velocity field in a work of Shibata and Taniguchi [68]. The initial condition of these works are computed by the puncture method, and the evolution are also performed using the moving-puncture method [101, 102]. Initial conditions computed by the excision method are evolved using the moving-puncture method in a work of Etienne et al. [103], and the BH spin is introduced in a later work of Etienne et al. [104]. All of these simulations are performed using the finite differentiation method both for geometric and hydrodynamic fields. The spectral method is applied for the geometric fields in a work of Duez et al. [105], where the merger of an equal-mass binary is performed. These simulations only tracks the relatively small number of orbits except for limited models, and long-term simulations of nonspinning BH–NS binaries are performed in a work of Shibata et al. [106] for a wide range of mass ratio to derive reliable gravitational waves using an AMR code, SACRA [107]. It should be noted that preliminary simulations of BH–NS mergers are also performed in the first paper of SACRA [107]. The effect of magnetic field[9] is first introduced in a work of Chawla et al. [108], and the fallback time of disrupted material is also estimated. The effect of the magnetic field on the late-time dynamics of the merger remnant is later investigated in a work of Etienne et al. [109]. The effect of the orientation of BH spins are investigated in a work of Foucart et al. [110], and simulations for a large mass ratio binary with a $10M_\odot$ BH are performed in another work of Foucart et al. [111], where the spectral method is applied for geometric fields. It is important that all of these simulations adopt a $\Gamma = 2$, ideal gas EOS. Exceptionally, ideal gas EOSs with $\Gamma \neq 2$ and the Shen EOS [112, 113] are adopted in the work of Duez et al. [114] for short-term simulations of a fixed mass ratio and a fixed spin parameter of the binary, whereas the microphysics is not incorporated to evolve the electron fraction.

To date, only limited number of simulations have been performed taking into account the nuclear-theory based EOS[10] [114]. In particular, we still do not understand the dependence of the merger process and resulting gravitational waveforms on the EOS of the NS in detail. The author of this thesis extensively performed systematic study of such dependence on the NS EOS for the first time in full general relativity, for a wide range of the mass ratio and the BH spin [117–119].

[9] Currently, there is no consistent method to compute quasiequilibrium states of magnetized compact binaries. The simulation for a magnetized compact binary is performed by superposing a magnetic field on a nonmagnetized initial condition, more or less artificially.

[10] We should also mention works done by Stephens et al. and East et al. (the same authors except for ordering) [115, 116], in which hyperbolic encounters of BH and NS are studied with the same EOS as that adopted in this thesis.

1.5 The Purpose of This Thesis and Convention

In this thesis, we study systematically the effect of the EOS on the merger process, on the properties of remnants, and on gravitational waves of the BH–NS binary merger. For this purpose, we perform fully general relativistic simulations of the BH–NS binary merger for a wide range of parameters, i.e., the NS mass, BH mass (or the mass ratio), BH spin, and NS EOS. Specifically, the NS EOS is modeled systematically by a piecewise polytropic EOS, which we describe in Sect. 2.4, for the first time in the study of the BH–NS binary merger. As for the BH spin, we only focus on the cases in which the BH spin is zero or (anti) aligned with the orbital angular momentum of the binary, and instead vary systematically the magnitude of the BH spin. The NS mass is chosen to be values around the canonical mass of observed binary NSs, $1.35M_{\odot}$ [28], and the mass ratio of the BH to the NS is also systematically varied from low masses to astrophysically realistic masses [120]. We focus mainly on the case in which tidal disruption of the NS occur, i.e., the mass ratio is small and/or the BH has a prograde[11] spin.

This thesis is organized as follows. In Chap. 2, we review properties of the NS associated with the EOS, and describe our model of the EOS adopted in this study. We describe our method of computing initial conditions and of dynamical simulations in Chaps. 3 and 4, respectively. Chapter 5 summarizes our diagnostics for dynamical simulations. We show the result of simulations for nonspinning BH–NS binary mergers in Chap. 6 to elucidate the effect of the NS EOS. The result of simulations for spinning BH–NS binary mergers are shown in Chap. 7, focusing effects of both the NS EOS and the BH spin. Finally, Chap. 8 is devoted to a summary and discussion.

Hereafter, we adopt the geometrical units in which $G = c = 1$, unless otherwise stated. Our convention of notation for physically important quantities of the binary is summarized in Table 1.1. The nondimensional spin parameter of the BH, total mass of the system at infinite separation, mass ratio, and compactness of the NS are defined as $a = S_{\rm BH}/M_{\rm BH}^2$, $m_0 = M_{\rm BH} + M_{\rm NS}$, $Q = M_{\rm BH}/M_{\rm NS}$, and $\mathscr{C} = M_{\rm NS}/R_{\rm NS}$, respectively.

Lowercase Latin $(i, j, \ldots)$ and Greek $(\mu, \nu, \ldots)$ indices denote spatial and spacetime components, respectively. Upper case Latin $(A, B, \ldots)$ indices denote components on a two surface, such as (θ, φ), unless otherwise stated. We use the spacetime metric signature $(-, +, +, +)$, for which the flat metric takes the form $\eta_{\mu\nu} = {\rm diag}(-1, 1, 1, 1)$ in Cartesian coordinates. We typically denote the spacetime, a spacelike hypersurface, and a spacelike two sphere by $\mathscr{M}$, Σ, and $\mathscr{S}$, respectively.

As we do not take electromagnetic fields and radiation into account in this study, geometric and hydrodynamic fields are always governed by the Einstein equations,

$$G_{\mu\nu} = 8\pi T_{\mu\nu}, \tag{1.32}$$

[11] In this thesis, "prograde" and "retrograde" spins mean the BH spins which are aligned and antialigned with the orbital angular momentum of the binary, respectively.

Table 1.1 Our convention of notation for physically important quantities of a binary

Symbol	
$M_{\rm irr}$	The irreducible mass of the BH
$S_{\rm BH}$	The magnitude of the BH spin angular momentum
$M_{\rm BH}$	The gravitational mass of the BH in isolation
$M_{\rm NS}$	The gravitational mass of the NS in isolation
$R_{\rm NS}$	The circumferential radius of the NS in isolation
M_0	The Arnowitt-Deser-Misner mass of the system
m_0	The total mass of the system at infinite separation
Ω	The orbital angular velocity
Ω_0	The orbital angular velocity of the initial configuration
Q	The mass ratio
$\mathscr{C}$	The compactness of the NS
a	The nondimensional spin parameter of the BH

the continuity equation of the baryon number or rest mass,

$$\nabla_\mu(\rho u^\mu) = 0, \tag{1.33}$$

and the local energy-momentum conservation equation,

$$\nabla_\alpha T^{\mu\alpha} = 0. \tag{1.34}$$

We always assume an ideal fluid for the matter field, for which the energy-momentum tensor is given by

$$T^{\mu\nu} = \rho h u^\mu u^\nu + P g^{\mu\nu}, \tag{1.35}$$

where h is the specific enthalpy defined by

$$h = 1 + \varepsilon + \frac{P}{\rho}, \tag{1.36}$$

and assume that the pressure P is determined from ρ and ε using the EOS of the form

$$P = P(\rho, \varepsilon). \tag{1.37}$$

Tables 1.2 and 1.3 summarizes our convention of notation for geometric and hydrodynamic variables, respectively. Notice that the energy density is given by

$$e = \rho(1 + \varepsilon) = \rho h - P. \tag{1.38}$$

Before closing this chapter, we summarize our mathematical convention [121, 122]. The Riemann tensor is defined by, for any 1-form ω_μ,

Table 1.2 Our convention of notation for geometric variables

Symbol	
$g_{\mu\nu}$	The spacetime metric
∇_μ	The covariant derivative associated with $g_{\mu\nu}$
${}^4\Gamma^\mu{}_{\alpha\beta}$	The Christoffel symbol of $g_{\mu\nu}$
${}^4R_{\mu\nu\alpha\beta}$	The Riemann tensor of $g_{\mu\nu}$
${}^4R_{\mu\nu}$	The Ricci tensor of $g_{\mu\nu}$
${}^4R$	The scalar curvature of $g_{\mu\nu}$
$\eta_{\mu\nu}$	The flat spacetime metric
n^μ	The timelike unit normal vector to a spacelike hypersurface Σ
γ_{ij}	The induced metric on a spacelike hypersurface Σ
D_i	The covariant derivative associated with γ_{ij}
$\Gamma^k{}_{ij}$	The Christoffel symbol of γ_{ij}
R_{ijkl}	The Riemann tensor of γ_{ij}
R_{ij}	The Ricci tensor of γ_{ij}
R	The scalar curvature of γ_{ij}
K_{ij}	The extrinsic curvature of γ_{ij}
K	The trace of K_{ij}
α	The lapse function
β^i	The shift vector
f_{ij}	The flat three metric
$\overset{\circ}{D}_i$	The covariant derivative associated with f_{ij}
$\overset{\circ}{\varepsilon}_{ijk}$	The Levi-Civita tensor associated with f_{ij}
s^i	The spacelike outward unit normal to a two sphere $\mathscr{S}$
q_{AB}	The induced two metric on a two sphere $\mathscr{S}$
$\mathscr{D}_A$	The covariant derivative associated with q_{AB}
ε_{AB}	The Levi-Civita tensor associated with q_{AB}
$\mathscr{R}$	The scalar curvature of q_{AB}
f_{AB}	The flat two metric on $\mathscr{S}$
l^μ	The outgoing null vector
k^μ	The ingoing null vector
$\Theta_{(l)}$	The expansion of l^μ

Some of them only appear in the Appendix

$$\nabla_\mu \nabla_\nu \omega_\alpha - \nabla_\nu \nabla_\mu \omega_\alpha = {}^4R_{\mu\nu\alpha}{}^\beta \omega_\beta. \tag{1.39}$$

Accordingly, for any vector v^μ, the relation

$$\nabla_\mu \nabla_\nu v^\alpha - \nabla_\nu \nabla_\mu v^\alpha = -{}^4R^\alpha_{\mu\nu\beta} v^\beta \tag{1.40}$$

holds as long as ∇_μ is a torsion-free derivative operator. Although some of them do not appear explicitly, it should be worthwhile to note that the four-, three-, and two-dimensional Levi-Civita tensors are defined by

Table 1.3 Our convention of notation for hydrodynamic variables

Symbol	
$T^{\mu\nu}$	The energy-momentum tensor
ρ	The baryon rest-mass density
u^{μ}	The four velocity of the fluid
P	The pressure
ε	The specific internal energy
h	The specific enthalpy
e	The energy density
w	The Lorentz factor of the fluid

$$\varepsilon_{\mu\nu\alpha\beta} = \sqrt{-g}(\mathrm{d}x^0)_\mu \wedge (\mathrm{d}x^1)_\nu \wedge (\mathrm{d}x^2)_\alpha \wedge (\mathrm{d}x^3)_\beta, \tag{1.41}$$

$$\varepsilon_{\mu\nu\alpha} = n^\beta \varepsilon_{\beta\mu\nu\alpha} = \sqrt{\gamma}(\mathrm{d}x^1)_\mu \wedge (\mathrm{d}x^2)_\nu \wedge (\mathrm{d}x^3)_\alpha, \tag{1.42}$$

$$\varepsilon_{\mu\nu} = n^\alpha s^\beta \varepsilon_{\alpha\beta\mu\nu} = \sqrt{q}(\mathrm{d}x^2)_\mu \wedge (\mathrm{d}x^3)_\nu, \tag{1.43}$$

because the integration is performed via these differential forms. The future directed volume element of a three dimensional spacelike hypersurface seen from the four dimensional spacetime is written by

$$\mathrm{d}V_\mu = -n_\mu \sqrt{\gamma}\mathrm{d}^3x. \tag{1.44}$$

The directed surface element of a two surface seen from the four dimensional spacetime is written by

$$\mathrm{d}S_{\mu\nu} = -(n_\mu s_\nu - s_\mu n_\nu)\sqrt{q}\mathrm{d}^2x, \tag{1.45}$$

or accordingly seen from the three-dimensional space by

$$\mathrm{d}S_i = s_i \sqrt{q}\mathrm{d}^2x. \tag{1.46}$$

The Gauss' theorem for the divergence of a vector v^μ and the Stokes' theorem for the divergence of an antisymmetric tensor $B^{\mu\nu}$,

$$\int_{\mathcal{M}} \nabla_\mu v^\mu \sqrt{-g}\mathrm{d}^4x = \oint_{\partial\mathcal{M}} v^\mu \mathrm{d}V_\mu, \quad \int_\Sigma \nabla_\nu B^{\mu\nu}\mathrm{d}V_\mu = \frac{1}{2}\oint_{\partial\Sigma} B^{\mu\nu}\mathrm{d}S_{\mu\nu}. \tag{1.47}$$

are sometimes used without notice.

References

1. J. Abadie et al., Nucl. Instrum. Methods. Phys. Res. Sect. A **624**, 223 (2010)
2. T. Accadia et al., Class. Quant. Grav. **28**, 025005 (2011)
3. R.A. Hulse, J.H. Taylor, Astrophys. J. **195**, L51 (1975)
4. K. Kuroda et al., Class. Quant. Grav. **27**, 084004 (2010)
5. B.S. Sathyaprakash, B.F. Schutz, Living Rev. Relativ. **12**, 2 (2009)
6. B. Abbott et al., Astrophys. J. **683**, L45 (2009)
7. B. Abbott et al., Astrophys. J. **706**, L203 (2009)
8. B. Abbott et al., Nature **460**, 990 (2009)
9. J. Abadie et al., Astrophys. J. **737**, 93 (2011)
10. L. Lindblom, Astrophys. J. **398**, 569 (1992)
11. M. Vallisneri, Phys. Rev. Lett. **84**, 3519 (2000)
12. J.S. Read, C. Markakis, M. Shibata, K. Uryū, J.D.E. Creighton, J.L. Friedman, Phys. Rev. D **79**, 124033 (2009)
13. V. Ferrari, L. Gualtieri, F. Pannarale, Phys. Rev. D **81**, 064026 (2010)
14. F. Pannarale, A. Tonita, L. Rezzolla, Astrophys. J. **727**, 95 (2011)
15. F. Pannarale, L. Rezzolla, F. Ohme, J.S. Read, Phys. Rev. D **84**, 104017 (2011)
16. C. Messenger, J. Read, Phys. Rev. Lett. **108**, 091101 (2012)
17. M. Punturo et al., Class. Quant. Grav. **27**, 194002 (2010)
18. P. Demorest, T. Pennucci, S. Ransom, M. Roberts, J. Hessels, Nature **467**, 1081 (2010)
19. V. Kalogera, K. Belczynski, C. Kim, R. O'Shaughnessy, B. Willems, Phys. Rep. **442**, 75 (2007)
20. L. Blanchet, Living Rev. Relativ. **9**, 4 (2006)
21. K. Hotokezaka, K. Kyutoku, H. Okawa, M. Shibata, K. Kiuchi, Phys. Rev. D **83**, 124008 (2011)
22. Y. Sekiguchi, K. Kiuchi, K. Kyutoku, M. Shibata, Phys. Rev. Lett. **107**, 051102 (2011)
23. Y. Sekiguchi, K. Kiuchi, K. Kyutoku, M. Shibata, Phys. Rev. Lett. **107**, 211101 (2011)
24. E. Berti, V. Cardoso, A.O. Starinets, Class. Quant. Grav. **26**, 163001 (2009)
25. S.L. Shapiro, S.A. Teukolsky, *Black Holes, White Dwarfs, and Neutron Stars: The Physics of Compact Objects* (Wiley, New York, 1983)
26. P.C. Peters, J. Mathews, Phys. Rev. **131**, 435 (1963)
27. P.C. Peters, Phys. Rev. **136**, B1224 (1964)
28. D.R. Lorimer, Living Rev. Relativ. **11**, 8 (2008)
29. K. Belczynski, V. Kalogera, F.A. Rasio, R.E. Taam, A. Zezas, T. Bulik, T.J. Maccarone, N. Ivanova, Astrophys. J. Suppl. **174**, 223 (2008)
30. J. Abadie et al., Class. Quant. Grav. **27**, 173001 (2010)
31. K. Belczynski, T. Bulik, C.L. Fryer, A. Ruiter, F. Valsecchi, J.S. Vink, J.R. Hurley, Astrophys. J. **714**, 1217 (2010)
32. K. Belczynski, R.E. Taam, V. Kalogera, F.A. Rasio, T. Bulik, Astrophys. J. **662**, 504 (2007)
33. R. O'Shaughnessy, V. Kalogera, K. Belczynski, Astrophys. J. **716**, 615 (2010)
34. R. Narayan, B. Paczyński, T. Piran, Astrophys. J. **395**, L83 (1992)
35. B.D. Metzger, E. Berger, Astrophys. J. **746**, 48 (2012)
36. T. Piran, Phys. Rep. **314**, 575 (1999)
37. T. Piran, Rev. Mod. Phys. **76**, 1143 (2004)
38. Y. Lithwick, R. Sari, Astrophys. J. **555**, 540 (2001)
39. E. Nakar, Phys. Rep. **442**, 166 (2007)
40. S.E. Woosley, Astrophys. J. **405**, 273 (1993)
41. E. Berger et al., Nature **438**, 988 (2005)
42. W. Fong, E. Berger, R. Chornock, N.R. Tanvir, A.J. Levan, J.F. Graham, A.S. Fruchter, A. Cucchiara, D.B. Fox, Astrophys. J. **730**, 26 (2011)
43. W.H. Lee, E. Ramirez-Ruiz, New J. Phys **9**, 17 (2007)
44. J. Goodman, A. Dar, S. Nussinov, Astrophys. J. **314**, L7 (1987)

45. D. Eichler, M. Livio, T. Piran, D.N. Schramm, Nature **340**, 126 (1989)
46. K. Kohri, S. Mineshige, Astrophys. J. **577**, 311 (2002)
47. R.D. Blandford, R.D. Znajek, Mon. Not. Roy. Astron. Soc. **179**, 433 (1977)
48. D. Macdonald, K.S. Thorne, Mon. Not. Roy. Astron. Soc. **198**, 345 (1982)
49. J.C. McKinney, Mon. Not. Roy. Astron. Soc. **368**, 1561 (2006)
50. J.M. Lattimer, D.N. Schramm, Astrophys. J. **192**, L145 (1974)
51. Y. Qian, G.J. Wasserburg, Phys. Rep. **442**, 237 (2007)
52. F.-K. Thielemann et al., Prog. Part. Nucl. Phys. **66**, 346 (2011)
53. S.E. Woosley, J.R. Wilson, G.J. Mathews, R.D. Hoffman, B.S. Meyer, Astrophys. J. **433**, 229 (1994)
54. L.X. Li, B. Paczyński, Astrophys. J. **507**, L59 (1998)
55. B.D. Metzger, G. Martínez-Pinedo, S. Darbha, E. Quataert, A. Arcones, D. Kasen, R. Thomas, P. Nugent, I.V. Panov, N.T. Zinner, Mon. Not. Roy. Astron. Soc. **406**, 2650 (2010)
56. J.P.A. Clark, D.M. Eardley, Astrophys. J. **215**, 311 (1977)
57. M. Shibata, K. Taniguchi, Living Rev. in Relativ. **14**, 6 (2011)
58. J.M. Bardeen, W.H. Press, S.A. Teukolsky, Astrophys. J. **178**, 347 (1972)
59. L.G. Fishbone, Astrophys. J. **185**, 43 (1973)
60. B. Mashhoon, Astrophys. J. **197**, 705 (1975)
61. J.A. Marck, Proc. Roy. Soc. Lond. **385**, 431 (1983)
62. M. Shibata, Prog. Theor. Phys. **96**, 917 (1996)
63. P. Wiggins, D. Lai, Astrophys. J. **532**, 530 (2000)
64. M. Ishii, M. Shibata, Y. Mino, Phys. Rev. D **71**, 044017 (2005)
65. V. Ferrari, L. Gualtieri, F. Pannarale, Class. Quant. Grav. **26**, 125004 (2009)
66. K. Taniguchi, T.W. Baumgarte, J.A. Faber, S.L. Shapiro, Phys. Rev. D **75**, 084005 (2007)
67. K. Taniguchi, T.W. Baumgarte, J.A. Faber, S.L. Shapiro, Phys. Rev. D **77**, 044003 (2008)
68. M. Shibata, K. Taniguchi, Phys. Rev. D **77**, 084015 (2008)
69. M. Miller, (2001)
70. S. Brandt, B. Brügmann, Phys. Rev. Lett. **78**, 3606 (1997)
71. J.W. York, in *Gravitational Radiation*, ed. by L. Smarr (Cambridge University Press, Cambridge, 1979)
72. T.W. Baumgarte, M.L. Skoge, S.L. Shapiro, Phys. Rev. D **70**, 064040 (2004)
73. K. Taniguchi, T.W. Baumgarte, J.A. Faber, S.L. Shapiro, Phys. Rev. D **72**, 044008 (2005)
74. P. Grandclément, J. Novak, Living Rev. Relativ. **12**, 1 (2009)
75. P. Grandclément, Phys. Rev. D **74**, 124002 (2006)
76. P. Grandclément, Phys. Rev. D **75**, 129903(E) (2007)
77. E. Gourgoulhon, P. Grandclément, S. Bonazzola, Phys. Rev. D **65**, 044020 (2002)
78. P. Grandclément, E. gourgoulhon, S. Bonazzola, Phys. Rev. D **65**, 044021 (2002)
79. K. Taniguchi, T.W. Baumgarte, J.A. Faber, S.L. Shapiro, Phys. Rev. D **74**, 041502(R) (2006)
80. G.B. Cook, Phys. Rev. D **65**, 084003 (2002)
81. G.B. Cook, H.P. Pfeiffer, Phys. Rev. D **70**, 104016 (2004)
82. F. Foucart, L.E. Kidder, H.P. Pfeiffer, S.A. Teukolsky, Phys. Rev. D **77**, 124051 (2008)
83. G. Lovelace, R. Owen, H.P. Pfeiffer, T. Chu, Phys. Rev. D **78**, 084017 (2008)
84. H.P. Pfeiffer, D.A. Brown, L.E. Kidder, L. Lindblom, G. Lovelace, M.A. Scheel, Class. Quant. Grav. **24**, S59 (2007)
85. K. Kyutoku, M. Shibata, K. Taniguchi, Phys. Rev. D **79**, 124018 (2009)
86. M. Shibata, K. Uryū, Phys. Rev. D **74**, 121503(R) (2006)
87. M. Shibata, K. Uryū, Class. Quant. Grav. **24**, S125 (2007)
88. W. Kluźniak, W.H. Lee, Astrophys. J. **494**, L53 (1998)
89. W.H. Lee, W. Kluźniak, Astrophys. J. **526**, 178 (1999)
90. H.T. Janka, T. Eberl, M. Ruffert, C.L. Fryer, Astrophys. J. **527**, L39 (1999)
91. J.M. Lattimer, F.D. Swesty, Nucl. Phys. A **535**, 331 (1991)
92. S. Rosswog, R. Speith, G.A. Wynn, Mon. Not. Roy. Astron. Soc. **351**, 1121 (2004)
93. S. Rosswog, Astrophys. J. **634**, 1202 (2005)
94. M. Ruffert, H.T. Janka, Astron. Astrophys. **514**, A66 (2010)

95. J. Faber, T.W. Baumgarte, S.L. Shapiro, K. Taniguchi, F.A. Rasio, Phys. Rev. D **73**, 024012 (2006)
96. J. Faber, T.W. Baumgarte, S.L. Shapiro, K. Taniguchi, Astrophys. J. **641**, L93 (2006)
97. J.R. Wilson, G.J. Mathews, Phys. Rev. Lett. **75**, 4161 (1995)
98. J.R. Wilson, G.J. Mathews, P. Marronetti, Phys. Rev. D **54**, 1317 (1996)
99. E. Rantsiou, S. Kobayashi, P. Laguna, F.A. Rasio, Astrophys. J. **680**, 1326 (2008)
100. C.F. Soupuerta, U. Sperhake, P. Laguna, Class. Quant. Grav. **23**, S579 (2006)
101. M. Campanelli, C.O. Lousto, P. Marronetti, Y. Zlochower, Phys. Rev. Lett. **96**, 111101 (2006)
102. J.G. Baker, J. Centrella, D.I. Choi, M. Koppitz, J. van Meter, Phys. Rev. Lett. **96**, 111102 (2006)
103. Z.B. Etienne, J.A. Faber, Y.T. Liu, S.L. Shapiro, K. Taniguchi, T.W. Baumgarte, Phys. Rev. D **77**, 084002 (2008)
104. Z.B. Etienne, Y.T. Liu, S.L. Shapiro, T.W. Baumgarte, Phys. Rev. D **79**, 044024 (2009)
105. M.D. Duez, F. Foucart, L.E. Kidder, H.P. Pfeiffer, M.A. Scheel, S.A. Teukolsky, Phys. Rev. D **78**, 104015 (2008)
106. M. Shibata, K. Kyutoku, T. Yamamoto, K. Taniguchi, Phys. Rev. D **79**, 044030 (2009)
107. T. Yamamoto, M. Shibata, K. Taniguchi, Phys. Rev. D **78**, 064054 (2008)
108. S. Chawla, M. Anderson, M. Besselman, L. Lehner, S.L. Liebling, P.M. Motl, D. Neilsen, Phys. Rev. Lett. **105**, 111101 (2010)
109. Z.B. Etienne, Y.T. Liu, V. Paschalidis, S.L. Shapiro, Phys. Rev. D **85**, 064029 (2012)
110. F. Foucart, M.D. Duez, L.E. Kidder, S.A. Teukolsky, Phys. Rev. D **83**, 024005 (2011)
111. F. Foucart, M.D. Duez, L.E. Kidder, M.A. Scheel, B. Szilágyi, S.A. Teukolsky, Phys. Rev. D **85**, 044015 (2012)
112. H. Shen, H. Toki, K. Oyamatsu, K. Sumiyoshi, Nucl. Phys. A **637**, 435 (1998)
113. H. Shen, H. Toki, K. Oyamatsu, K. Sumiyoshi, Prog. Theor. Phys. **100**, 1013 (1998)
114. M.D. Duez, F. Foucart, L.E. Kidder, C.D. Ott, S.A. Teukolsky, Class. Quant. Grav. **27**, 114106 (2010)
115. B.C. Stephens, W.E. East, F. Pretorius, Astrophys. J. **737**, L5 (2011)
116. W.E. East, F. Pretorius, B.C. Stephens, Phys. Rev. D **85**, 124009 (2012)
117. K. Kyutoku, M. Shibata, K. Taniguchi, Phys. Rev. D **82**, 044049 (2010)
118. K. Kyutoku, M. Shibata, K. Taniguchi, Phys. Rev. D **84**, 049902(E) (2011)
119. K. Kyutoku, H. Okawa, M. Shibata, K. Taniguchi, Phys. Rev. D **84**, 064018 (2011)
120. J.E. McClintock, R.A. Remillard, in *Compact Stellar X-ray Sources*, ed. by W.H.G. Lewin, M. van der Klis (Cambridge University Press, Cambridge, 2006)
121. R.M. Wald, *General Relativity* (The university of Chicago press, 1984)
122. E. Poisson, *A Relativist's Toolkit: The Mathematics of Black-Hole Mechanics* (Cambridge university press, Cambridge, 2004)

References

Chapter 2
Equations of State of Neutron Star Matter

The purpose of this thesis is to investigate dependence of gravitational waves on the NS EOS, aiming for determining or at least constraining the EOS at high density above the nuclear saturation density $\rho_{\rm sat} \approx 2.7 \times 10^{14}\,{\rm g\,cm^{-3}}$ (or $n_{\rm sat} \approx 0.16\,{\rm fm^{-3}}$ in terms of the baryon number density) by gravitational-wave observation. Here, we mean "cold, zero-temperature EOS" by "EOS," because the finite-temperature effect is believed to have a negligible role in determining the property of an isolated NS and a NS just before the merger [1]. In this chapter, we review relations between the NS properties and the NS EOS from a theoretical point of view, as well as describe how the NS EOS is constrained by current astrophysical observation based on electromagnetic radiation. Finally, we describe our model of the cold, zero-temperature EOS, i.e., the piecewise polytropic EOS [2], adopted in this thesis.

2.1 Formation and Cooling of (Proto-)Neutron Stars

The NS is believed to be formed after the core-collapse (Type-II) supernova of a star with 8–25$M_\odot$ at the zero-age main sequence, where the precise mass range is uncertain. The reason for this uncertainty is, on the one hand, associated with theoretical incompleteness in the modeling of supernova explosions. On the other hand, the rotation and metalicity of the progenitor change the explosion characteristics, and therefore the remnant object is not uniquely determined by the progenitor mass due to a physical reason. At any rate, the NS is thought to be formed from the progenitor's iron core with the mass comparable to the Chandrasekhar mass, $\lesssim 1.4\,M_\odot$, where the exact value depends on the lepton fraction of the core. Here, we define the lepton fraction to be a ratio of the lepton number density to the baryon number density, as well as for the fraction of other particles. A newly-born NS at the core bounce is a lepton-rich object with a large amount of trapped neutrinos due to a nucleonic coherent scattering [3], and is surrounded by a shocked, high-entropy mantle [4, 5]. The nascent NS is usually called the "proto-neutron star (PNS)," because the lepton

K. Kyutoku, *The Black Hole–Neutron Star Binary Merger in Full General Relativity*, Springer Theses, DOI: 10.1007/978-4-431-54201-8_2, © Springer Japan 2013

fraction is so large that neutrons are not the distinctly dominant constituent. If the shock wave blows the envelope off and then accretion halts (a successful supernova), the mantle emits the neutrino extensively and collapses with increasing its temperature up to ~50 MeV. The PNS structure, such as the rest-mass density distribution, is already close to that of a NS when the mantle collapses. Subsequently, neutrinos in the PNS core escape on a diffusion time scale ~10–15 s, and neutronize the PNS into a NS via the reduction of the lepton fraction and the inverse β-decay. The deleptonization at this stage results in a PNS heating up to ~30 MeV, because the escaping neutrinos warm the PNS material by entropy generation during their diffusion, which is often compared to the Joule heating. The heat flux from the collapsed mantle also helps the warming of the interior. If some exotic particles such as hyperons emerge as a result of the deleptonization, the PNS may collapse to a BH due to a possible instability. Care must be taken for the fact that this evolutionary history neglects the convection. The study of the PNS convection is in progress mainly from the perspective of the supernova explosion, for which the PNS convection is expected to work as a mechanism to increase the neutrino luminosity (see, e.g., [6, 7]), whereas it may not be important for the purpose of this section.

After ~1 min of the core bounce, the mean free path of a neutrino becomes comparable to the NS radius, and the NS becomes transparent to neutrinos so that the NS starts to cool down rapidly (see [8, 9] and references therein for reviews). The neutrino emissivity is high in the core region of the NS, and the crust region[1] serves as a heat reservoir at the early epoch. As a result of the thermal transport from the crust to the core by the electron conduction, the NS gradually establishes an isothermal structure in the thermal evolution time scale of the crust, ~10–100 year [10]. This time scale depends on the NS structure itself, and also on the thermal conductivity and heat capacity, which are determined by microphysics such as the crustal neutron superfluidity. After the isothermal structure is established, the cooling time scale of the NS depends on the available cooling channel. One scenario is the so-called enhanced cooling scenario, in which the direct URCA process,

$$n \to p + e^- + \bar{\nu}_e, \; p \to n + e^+ + \nu_e, \tag{2.1}$$

is the main agent of neutrino emission. Although this process is very efficient to radiate the thermal energy, it is not believed to occur for a long time, because it is difficult to conserve energy and momentum simultaneously in these reactions due to by far the large Fermi momentum of the neutron [1]. The conservations can be satisfied only when the proton fraction is larger than ≈1/9 for such low temperature [11]. This does not seem to be the case for the NS, for which the proton fraction is estimated to be a few percent at the nuclear saturation density by the β-equilibrium condition [12]. It should be noted that, however, the direct URCA

[1] The definition of the crust and core of the NS is not decisive. In this section, we assume that the crust and core is approximately divided by the rest-mass density $\simeq \rho_{\rm sat}/2$ at which nuclei begins to merge into a huge nucleus. Our discussion does not depend on the precise value of the dividing rest-mass density, as far as the crust region is reasonably thin.

process becomes possible if exotic particles are present. Another promising scenario is the so-called standard scenario, in which the modified URCA process,

$$n + (n, p) \to p + (n, p) + e^- + \bar{\nu}_e, \; p + (n, p) \to n + (n, p) + e^+ + \nu_e, \quad (2.2)$$

is the agent of neutrino emission. This process satisfies the energy and momentum conservations without difficulty thanks to the bystander particle. The deficit of this scenario is the cooling rate is strongly suppressed by a factor of $(kT/\mu_n)^2$ compared to the direct URCA process. Here, k and μ_n are the Boltzmann constant[2] and the chemical potential of the neutron, respectively. In a typical situation, the chemical potential (or the Fermi energy) of the NS matter is $\gtrsim$50 MeV around the nuclear saturation density and the NS temperature is ~0.1–1 MeV at this stage [13], the cooling rate is reduced by a factor of $\sim 10^4$–10^5.

Figure 2.1 shows theoretical cooling curves for the two scenarios described above. Observational data obtained by thermal X-ray emission are also shown, and these observations suggest that the standard cooling scenario is close to the realistic one.

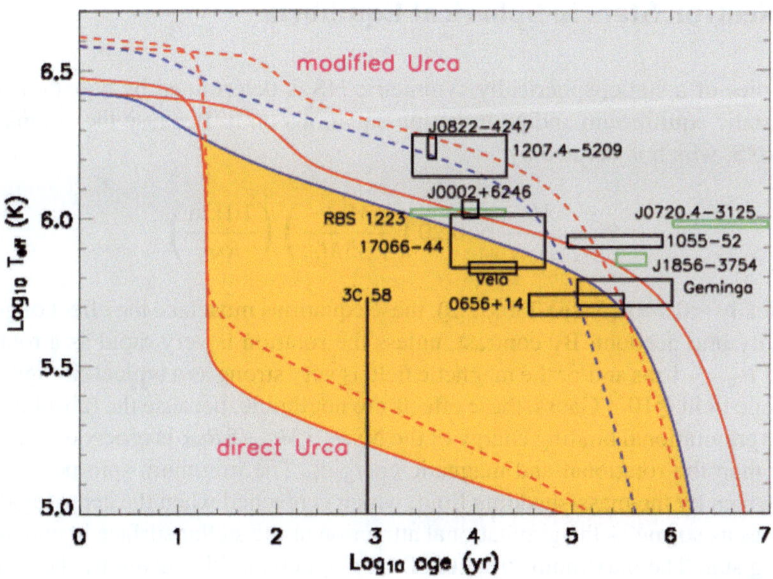

Fig. 2.1 Theoretical cooling curves for the standard (modified URCA) and enhanced cooling (direct URCA) scenario for a $1.4M_\odot$ NS. *Upper* four curves are for the standard scenario, and the *lower* two curves are for the enhanced cooling scenario. The solid and dashed curves correspond to models which adopt Fe and H envelopes, respectively. The *red* curves include the superfluidity and the *blue* curves do not. The *black* boxes show observational data obtained by the thermal X-ray emission, and the *green* boxes show the data for which the thermal optical emission is also reported. The age is determined by the pulsar spin down or the proper motion of the star [9]. This figure is taken from [14]

[2] $k = 1.38 \times 10^{-16}$ erg K^{-1}.

It is also interesting that the existence of the neutron superfluidity in the core region is strongly suggested recently (at the end of 2011) by observing the rapid drop of the surface temperature, which may be ascribed to breaking and formation of neutron Cooper pairs, of the NS in Cassiopeia A by *Chandra* [15–17]. From the perspective of the binary merger, a typical cooling time scale is orders of magnitude shorter the typical time scale of the BH–NS binary merger,[3] which is estimated to be $\sim$1 Gyr by the population synthesis [18, 19]. For the purpose of this study, it should again be noted that the typical Fermi energy of a zero temperature NS matter at the nuclear saturation density is $\gtrsim 50$ MeV, and the interior temperature is assumed to be only a factor of $\sim$100 larger than the effective surface temperature shown in Fig. 2.1 [20]. Therefore, the matter inside the NS in the late inspiral phase is safely believed to be well-approximated by a zero-temperature nuclear matter. Hence, we employ a cold EOS, for which the rest-mass density, ρ, determines all other thermodynamical quantities for calculating both the quasiequilibrium state of the BH–NS binary and the coalescence of the BH–NS binary up to the merger.

2.2 Neutron Stars in Spherical Equilibria

Properties of a static, spherically symmetric NS is determined by equations of the hydrostatic equilibrium and perturbation equations on it. Because the compactness of the NS, which is defined by

$$\mathscr{C} \equiv \frac{M_{\rm NS}}{R_{\rm NS}} \simeq 0.20 \left(\frac{M_{\rm NS}}{1.35 M_\odot} \right) \left(\frac{10\,{\rm km}}{R_{\rm NS}} \right), \tag{2.3}$$

is typically $\approx$0.1–0.25 (see Table 2.2), these equations must take the effect of general relativity into account. By contrast, unless the rotation is very rapid as a rotational period $P_{\rm rot} \lesssim 1$ ms and/or the magnetic field is very strong as a typical strength of the magnetic field $\gtrsim 10^{17}$ Gauss, these effects are negligible, because the (absolute value of the) gravitational binding energy of the NS $\approx 3M_{\rm NS}^2/5R_{\rm NS}$ is orders of magnitude larger than the rotational and magnetic energies. The minimum spin period of the NS is given by the mass-shedding limit, which is reached when the centrifugal force becomes as strong as the gravitational attraction at the stellar surface in context of a rotating star. The maximum strength of the magnetic field is given by the so-called virial limit, [21] which is reached when the scalar virial theorem is satisfied by the competition of the gravitational and magnetic energies. For properties of rotating NSs, see [22] and references therein for reviews. Properties of NSs with purely poloidal [23, 24] or purely toroidal [25, 26] magnetic fields are investigated, but currently there exists no equilibrium model of NSs with both poloidal and toroidal magnetic fields in full general relativity (but see [27–29]).

[3] This is also the case for the binary NSs.

2.2.1 The Tolman-Oppenheimer-Volkoff Equation

The structure of a static, spherically symmetric star is described by the Tolman-Oppenheimer-Volkoff (TOV) equation [30, 31] in general relativity.[4] In a static, spherically symmetric spacetime, the line element is written as

$$\mathrm{d}s^2 = -\mathrm{e}^{\nu(r)}\mathrm{d}t^2 + \mathrm{e}^{\lambda(r)}\mathrm{d}r^2 + r^2(\mathrm{d}\theta^2 + \sin^2\theta \mathrm{d}\varphi^2) \tag{2.4}$$

without loss of generality, and nonzero components of the four velocity and the energy-momentum tensor are given by

$$u^t = \mathrm{e}^{-\nu(r)/2},\ T_t{}^t = -e,\ T_r{}^r = T_\theta{}^\theta = T_\varphi{}^\varphi = P. \tag{2.5}$$

(Notice that ν and λ are not the components of tensorial quantities, but functions in this section. We expect that the confusion may not arise in this context.) From r component of the local energy-momentum conservation equation and *tt*, *tr*, and *rr* components of the Einstein equations, equations governing the hydrostatic equilibrium is derived as

$$\frac{\mathrm{d}P}{\mathrm{d}r} = -\frac{(e+P)(m+4\pi P r^3)}{r(r-2m)}, \tag{2.6}$$

$$\frac{\mathrm{d}\nu}{\mathrm{d}r} = -\frac{2}{e+P}\frac{\mathrm{d}P}{\mathrm{d}r}, \tag{2.7}$$

$$\mathrm{e}^{\lambda(r)} = \left(1 - \frac{2m}{r}\right)^{-1}, \tag{2.8}$$

$$m(r) \equiv 4\pi \int_0^r e(r')r'^2 dr', \tag{2.9}$$

where (2.6) is usually called the TOV equation. Hereafter, we focus on the NS, whereas discussion below also holds for generic stars such as white dwarfs.

When the EOS of the form $e = e(P)$ for the NS matter is specified, these equations are solved numerically (some analytic solutions are also known [30, 33, 34]) from the stellar center $r = 0$ to the stellar surface $r = R_{\mathrm{NS}}$, which is defined as the location where the pressure vanishes. It should be noted that the coordinate radius of the NS, R_{NS}, obtained by solving the TOV equation is the circumferential radius, and has a physical meaning. Initial conditions are

$$P = P_{\mathrm{c}},\ \nu = \nu_0,\ m = 0, \tag{2.10}$$

[4] In Newtonian gravity, a spherical equilibrium for a polytropic EOS is described by the Lane-Emden equation. The TOV equation with the (energy-)polytropic EOS is sometimes called the relativistic Lane-Emden equation [32], whereas it is not as useful as the Newtonian one.

where P_c is the value of the pressure at the stellar center and ν_0 is an arbitrary constant. The precise value of $\nu(r)$ is changed by adding a constant so that $\nu(R_{NS}) = 1 - 2M_{NS}/R_{NS}$ is satisfied, where $M_{NS} \equiv m(R_{NS})$ is the gravitational mass of the NS, after solving the equations. For a given EOS, a sequence of the spherical hydrostatic equilibrium is obtained as a one-parameter family of P_c, or is usually reinterpreted as the family of the central rest-mass density ρ_c. The relation between the gravitational mass, M_{NS}, and the circumferential radius, R_{NS}, of the spherical equilibrium sequence is called the *M–R* relation of the NS.

2.2.2 The Tidal Love Number and Deformability

The $l = 2$ tidal deformability $\lambda \equiv \lambda_2$ plays the most important role in the evolution of a compact binary among the quantities associated with the EOS [35]. The tidal effect on the orbital evolution comes into play even at the Newtonian order, and dominates the dynamics of a compact binary in the late inspiral phase [36, 37]. The tidal deformability represents the strength of the tidal effect on a generic star in a quantitative manner. Although we do not investigate dependence of gravitational waves and merger remnants on the tidal deformability in this thesis, it has considerable significance to describe it for future development of gravitational-wave templates [38]. The tidal deformability describes the response of a multipole moment of the star to the tidal force, and the l-mode tidal deformability λ_l with dimension [$\mathrm{g\,cm}^{2l-2}\,\mathrm{s}^2$] is defined via the nondimensional tidal Love number k_l as

$$\lambda_l = \frac{2R_{NS}^{2l+1}}{(2l-1)!!G}k_l, \tag{2.11}$$

where we inserted G for clarity. More conveniently, nondimensional tidal deformability

$$\Lambda_l \equiv G\lambda_l \left(\frac{c^2}{GM_{NS}}\right)^{2l+1} = \frac{2}{(2l-1)!!}k_l \left(\frac{c^2 R_{NS}}{GM_{NS}}\right)^{2l+1}, \tag{2.12}$$

where we inserted G and c for clarity, is sometimes introduced [38]. The tidal Love number characterizes the structure of a stellar configuration, which strongly depends on the EOS for a star with given values of mass and radius, in a nondimensional manner and in a different manner than the radius. In Newtonian gravity, tidal force tensor due to the external gravitational potential Φ_{ext} is defined by

$$\mathscr{E}_{ij} \equiv \frac{\partial^2 \Phi_{ext}}{\partial x^i \partial x^j}, \tag{2.13}$$

and the mass quadrupole moment induced by the tidal force field is defined by (1.4),

$$\mathcal{Q}_{ij} \equiv \int \rho \left(x_i x_j - \frac{1}{3} r^2 f_{ij} \right) \mathrm{d}^3 x. \tag{2.14}$$

Assuming that the background configuration is spherically symmetric, the tidal deformability relates these two quantities by

$$\mathcal{Q}_{ij} = -\lambda \mathcal{E}_{ij} \tag{2.15}$$

to a linear order of $\mathcal{E}_{ij}$. Although the computation of the tidal Love number, k_l, in Newtonian gravity[5] is found in the literature, e.g., [39, 40], the tidal Love number of the NS has to be evaluated in general relativity as well as the mass and radius due to its strong self-gravity.

General relativistic computation of the tidal Love number is performed by solving the $l = 2$, static, and even-parity perturbation equation on the background stellar configuration described by the TOV equation. Whereas precise formulations for this computation is developed in [41, 42] in different coordinates, we describe the original formulation of [43, 44] here for simplicity. Taking the Regge-Wheeler gauge [45], perturbations on the metric and hydrodynamic variables are both decomposed into modes associated with a spherical harmonics, $Y^{lm}(\theta, \varphi)$, relying on the background spherical symmetry. Hereafter, we only focus on the $(l, m) = (2, 0)$ mode, because the dominant role is played by the $l = 2$ mode and the perturbation equation does not depend on m. The static, even-parity perturbation on the metric is written using three functions $H_0(r)$, $H_2(r)$, and $K(r)$ by

$$g_{\alpha\beta} = g^{(0)}_{\alpha\beta} + h_{\alpha\beta}, \tag{2.16}$$

$$h_{\alpha\beta} = \mathrm{diag}\left[-\mathrm{e}^{\nu(r)} H_0(r), \mathrm{e}^{\lambda(r)} H_2(r), r^2 K(r), r^2 \sin^2\theta K(r) \right]_{\alpha\beta} \times Y^{20}(\theta, \varphi), \tag{2.17}$$

where $g^{(0)}_{\alpha\beta}$ is the solution of the TOV equation. Nonzero components of the perturbation on the four velocity and the energy-momentum tensor are

$$\delta u^t = -\mathrm{e}^{-\nu(r)/2} \frac{H_0(r)}{2}, \ \delta T_t{}^t = -\delta e = -\left(\frac{\mathrm{d}P}{\mathrm{d}e} \right)^{-1} \delta P,$$
$$\delta T_r{}^r = \delta T_\theta{}^\theta = \delta T_\varphi{}^\varphi = \delta P. \tag{2.18}$$

The perturbation equations are derived in [46] for general perturbations on a spherically symmetric background. Specifically in this case, the $(\theta\theta - \varphi\varphi)$ component of the perturbed Einstein equations shows $H \equiv H_0 = -H_2$. Using $r\theta$ and $(\theta\theta + \varphi\varphi)$ components to eliminate K and δP, the master equation of the perturbation H is obtained by the $(tt + \varphi\varphi)$ component as

[5] In Newtonian gravity, the tidal Love number is computed by the Clairaut-Ladau equation.

$$\frac{d^2H}{dr^2} + \frac{dH}{dr}\left\{\frac{2}{r} + e^{\lambda}\left[\frac{2m}{r^2} + 4\pi r(P-e)\right]\right\}$$
$$+ H\left[-\frac{6e^{\lambda}}{r^2} + 4\pi e^{\lambda}\left(5e + 9P + \frac{e+P}{dP/de}\right) - \left(\frac{d\nu}{dr}\right)^2\right] = 0, \tag{2.19}$$

and is rewritten as a first-order differential equation for the logarithmic derivative,

$$y(r) \equiv \frac{r}{H(r)}\frac{dH(r)}{dr}, \tag{2.20}$$

by [47]

$$\frac{dy}{dr} + \frac{y^2}{r} + ye^{\lambda}\left[\frac{1}{r} + 4\pi r(P-e)\right]$$
$$+ r\left[-\frac{6e^{\lambda}}{r^2} + 4\pi e^{\lambda}\left(5e + 9P + \frac{e+P}{dP/de}\right) - \left(\frac{d\nu}{dr}\right)^2\right] = 0. \tag{2.21}$$

This equation is solved from the stellar center to obtain $y(R_{NS})$ with initial conditions obtained from the regularity condition,

$$H = A_2 r^2, \ \frac{dH}{dr} = 2A_2 r, \ y = 2, \tag{2.22}$$

where the value of constant A_2 is arbitrary, because the problem is the linear perturbation. The choice of A_2 does not influence the value of y and hence the tidal Love number, as is expected.

Outside the star, homogeneous solutions to this equation are given analytically in terms of the associated Legendre polynomials of order $m = 2$, ${P_2}^2$ and ${Q_2}^2$, as

$$H(r) = c_1 {Q_2}^2\left(\frac{r}{M} - 1\right) + c_2 {P_2}^2\left(\frac{r}{M} - 1\right), \tag{2.23}$$

where c_1 and c_2 are constants, and this homogeneous solution asymptotically approaches

$$H(r \to \infty) \to \frac{8}{5}\left(\frac{M}{r}\right)^3 c_1 + 3\left(\frac{r}{M}\right)^2 c_2 \tag{2.24}$$

at large r. Meanwhile, in a local asymptotic frame of the NS, the metric coefficient[6] which corresponds to the gravitational potential at large r is given by [48, 49]

[6] The metric coefficients depend on the gauge choice, and actually the expression shown here is derived not in the Regge-Wheeler gauge but in the de Donder gauge (harmonic coordinate). More rigorous discussion of the gauge issues are presented in [41, 42].

$$-\frac{1+g_{tt}}{2} \to -\frac{M_{\rm NS}}{r} - \frac{3}{2r^3}\mathscr{Q}_{ij}\hat{x}^i\hat{x}^j + \frac{r^2}{2}\mathscr{E}_{ij}\hat{x}^i\hat{x}^j, \tag{2.25}$$

where $\hat{x}^i \equiv x^i/r$. Assuming that the quadrupole moment is induced by the tidal field via the tidal deformability λ as $\mathscr{Q}_{ij} = -\lambda\mathscr{E}_{ij}$, the constants in the homogeneous solution $H(r)$ becomes

$$c_1 = \frac{15}{8M_{\rm NS}^3}\lambda\mathscr{E}, \; c_2 = \frac{M_{\rm NS}^2}{3}\mathscr{E}, \tag{2.26}$$

where $\mathscr{E} \equiv \mathscr{E}_{ij}\hat{x}^i\hat{x}^j$. We also define the relativistic tidal Love number from λ as $k_2 = 3\lambda/2R_{\rm NS}^5$. Requiring the value and derivative of $H(r)$ to be continuous at the stellar surface, we finally obtain the value of the tidal Love number,

$$\begin{aligned} k_2 \equiv & \frac{8\mathscr{C}^5}{5}(1-2\mathscr{C})^2[2+2\mathscr{C}(y-1)-y] \\ & \times \Big\{ 2\mathscr{C}\left[6-3y+3\mathscr{C}(5y-8)\right] + 4\mathscr{C}^3[13-11y+\mathscr{C}(3y-2)+2\mathscr{C}^2(1+y)] \\ & + 3(1-2\mathscr{C})^2[2-y+2\mathscr{C}(y-1)]\ln(1-2\mathscr{C}) \Big\}^{-1}. \end{aligned} \tag{2.27}$$

In this expression, the prefactor $(1-2\mathscr{C})^2$ makes the tidal Love number rapidly decreasing function of the compactness, and the tidal Love number of a Schwarzschild BH is known to be zero [41]. The tidal Love numbers for other l modes are obtained by changing 6 in (2.21) to $l(l+1)$, changing 2 in (2.22) to l, and homogeneous solutions in the vacuum, (2.23), to the corresponding associated Legendre polynomials of degree l and order $m = 2$ assuming appropriate behavior at $r \to \infty$.

Before closing this section, we note that k_l should be called the electric Love number in general relativity, because the NS has the magnetic and shape Love numbers in addition [42]. In Newtonian gravity, the electric tidal Love number is also called the apsidal constant or the second tidal Love number, and the shape Love number corresponds to the first tidal Love number [40].

2.3 Current Constraints on the Equation of State

The NS EOS is highly uncertain so far, even at zero temperature. The theoretical calculation of the NS EOS must handle the many-body problem of strongly interacting particles, such as nucleons and mesons. Several computations are performed incorporating different particles and interactions, and adopting different computational methods of the many-body problem, such as the variational method and the relativistic mean field theory (see candidate EOSs for, e.g., [2] and references therein). These computations predict fairly different EOSs, and therefore the NS EOS is far from convergent from the theoretical point of view. Although the lattice quantum chromodynamics are now trying to perform ab initio computations of the NS EOS, there remains many difficulties to obtain the realistic EOS of the NS. It is also difficult to investigate properties of the NS matter by terrestrial experiments, because the NS matter is cold, high density, and neutron-rich. All of these properties make the experimental investigation difficult, and our experimental knowledge of the NS matter is quite limited.

Another way to know the NS EOS is to observe properties of actual NSs in our Universe. Although both accurate observation and an accurate modeling of the NS are necessary, the astronomical observation provides invaluable information of the NS and the NS EOS. In this section, we review the current constraints of the EOS obtained by astronomical observation of NSs. (See [13] for a thorough review of this issue.)

2.3.1 The Maximum Mass

The existence of a NS maximum mass is a remarkable feature of general relativity, and the NS maximum mass depends strongly on the NS EOS at high density. In particular, inclusion of the strong interaction in addition to the degenerate pressure of neutrons is essential to explain the observed NS mass [1]. Similarly, observation of a massive NS can rule out the EOS candidate when the maximum mass of the NS predicted by the EOS candidate is smaller than the observed mass of the NS.

We also address another importance of the NS maximum mass as a discriminator of the BH and NS. Whereas it is difficult to obtain a tight theoretical bound on the NS maximum mass independent of the EOS, it is of interest because the theoretical bound enables us to observationally distinguish BHs from NSs without precise knowledge of the NS EOS. A good or bad famous theoretical bound, $M_{\rm max} = 3.2 M_\odot$, is obtained in [50] by the variational method (see also [51]). The assumption there is that the EOS below a matching density $e_{\rm match} = 4.6 \times 10^{14}\ {\rm g/cm^3}$ is that of a free, degenerate neutron gas, and the EOS above the matching density is chosen so that the sound velocity is equal to the speed of light, $P = e - e_{\rm match} + P_{\rm match}$, where $P_{\rm match}$ is the pressure of the degenerate neutron gas at the matching density. Later in [51, 52], which tackle the problem by separating the core and envelope of the NS according to assumed EOSs, dependence of the maximum mass on the matching density is found to be

$$M_{\rm max}(e_{\rm match}) = 6.8 \left(\frac{e_{\rm match}}{10^{14}\ {\rm g\,cm^{-3}}} \right)^{-1/2} M_\odot, \tag{2.28}$$

when the causality is imposed on the EOS. It is also shown that this value depends only very weakly on the EOS below $e_{\rm match}$, as far as the low-density EOS is chosen to be that calculated from nuclear theory and the sound velocity is sufficiently smaller than the speed of light. From this dependence of the matching density, it is evident that the firm knowledge of the NS EOS up to higher density improves the theoretical bound of the NS maximum mass.

At the end of 2011, the most reliable.[7] constraint on the NS EOS is obtained by a radio pulsar-timing observation of a NS-white dwarf binary PSR J1614-2230, which is observed as a millisecond pulsar [53]. In addition to its fairly relativistic character, a remarkable feature of this binary is its very large inclination angle of the orbital plane as 89.17 ± 0.02 degree, and therefore the binary configuration is approximately edge-on. This geometry allows us to accurately measure the Shapiro time delay [57] of the pulse from the NS caused by the white dwarf companion when the NS is behind the white dwarf. The mass of the NS is determined to be $1.97 \pm 0.04 M_{\odot}$ within a 1-σ error, and therefore the EOS which cannot support $1.93 M_{\odot}$ is now considered to be ruled out.

Figure 2.2 shows this maximum mass constraint with the *M–R* relation of the NS for several nuclear-theory-based EOSs, as well as some other constraints (see below). Many of EOSs including exotic particles such as hyperons (actually, all EOSs plotted in Fig. 2.2) are excluded by this maximum mass limit. The reason for this is that the emergence of new particles unavoidably reduces the degeneracy of neutrons, and results in a softening of the EOS and a smaller value of the maximum mass. This

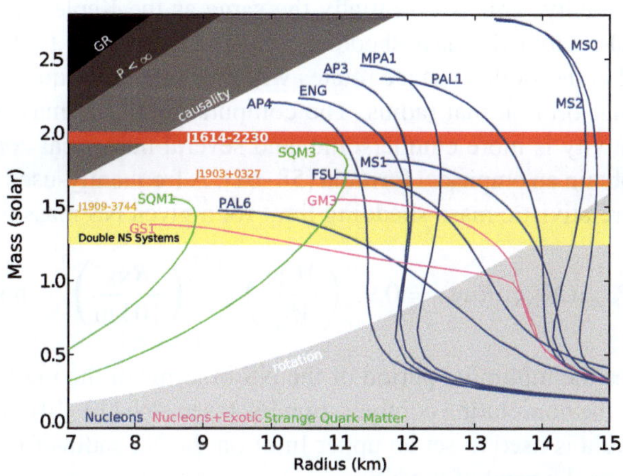

Fig. 2.2 The summary of constraints on the NS EOS. The *curves* labeled by EOS names are the relation between the mass and circumferential radius of a spherical NS for each EOS. The EOS which does not go above the band labeled by "J1614-2230," of which the NS mass is $1.97 \pm 0.04 M_{\odot}$ [53], is considered to be ruled out. The *top left* regions above "GR," "$P < \infty$," and "causality" is excluded theoretically by the condition in which the horizon does not exist ($R_{\rm NS} = 2M_{\rm NS}$), the pressure at the stellar center is finite ($R_{\rm NS} = 9M_{\rm NS}/4$) [54], and the EOS is causal, i.e., the sound velocity does not exceed the speed of light ($R_{\rm NS} \approx 2.9 M_{\rm NS}$), respectively. The *right bottom* region below "rotation" is excluded by the observation of J1748-2446ad, which emits the pulse at 714 Hz [55], so that the NS radius is not larger than the general relativistic mass-shedding limit [14]. Masses of some other NSs are also shown. This figure is taken from [53]

[7] Another candidate of the NS with the maximum mass is the black widow pulsar, PSR B1957+20. It is reported that the NS mass may be $2.40 \pm 0.12 M_{\odot}$ [56], but the reliability of the value is questioned.

does not immediately mean that the exotic particles do not exist, however, because the addition of repulsive interactions such as ones mediated by vector bosons will raise the maximum mass above the constraint, and will save EOSs including exotic particles.

2.3.2 The Minimum Rotational Period

The rotational period of the NS is limited from below by the mass-shedding limit, below which the centrifugal force at the stellar surface surpasses the NS self-gravity, and therefore the NS cannot maintain itself. Whereas it is difficult to take the change in the radius due to the rotational deformation into account, the mass-shedding limit is written using the NS mass, $M_{\rm NS}$, and the equatorial radius of the rotating NS, $R_{\rm eq}$, as

$$P_{\rm shed}(M_{\rm NS}, R_{\rm eq}) = 2\pi\sqrt{\frac{R_{\rm eq}^3}{M_{\rm NS}}}, \tag{2.29}$$

in Newtonian gravity. This is essentially the same as the Keplerian orbital period, and the general-relativistic mass-shedding limit is determined by the condition in which material at the stellar surface rotates with the angular frequency of a test particle in a circular orbit at that radius. The computation of the mass-shedding limit in general relativity is more cumbersome, and several numerical computations are performed to obtain an empirical formula [58–61]. A frequently-used fitting formula of the general-relativistic mass-shedding limit for a given NS mass is

$$P_{\rm shed}(M_{\rm NS}, R_{\rm NS}) = 0.93\left(\frac{M_{\rm NS}}{M_\odot}\right)^{-1/2}\left(\frac{R_{\rm NS}}{10\,{\rm km}}\right)^{3/2}\ {\rm ms}, \tag{2.30}$$

which describes the minimum period of the NS in terms of the mass, $M_{\rm NS}$, and the radius, $R_{\rm NS}$, of the nonrotating configuration [62] (see also [13, 14]). In practice, this empirical formula is used to set an upper limit on the NS radius from the observed mass and rotational period of the NS.

At the end of 2011, the most rapidly rotating NS known is PSR J1748-2466ad, which emits the pulse at 714 Hz [55]. The right bottom region of Fig. 2.2 depicts the region excluded by combining this observation and (2.30).[8] It should be noted that the radius constraint for a given value of $P_{\rm rot}$ becomes more severe for a smaller value of $M_{\rm NS}$. At any rate, the constraint given by J1748-2466ad alone is not so restrictive. Another candidate of the most rapidly rotating NS is an X-ray transient XTE J1739-285, of which the X-ray burst is observed to oscillate at 1122 Hz [63]. However, it is uncertain whether this oscillation frequency really corresponds to the NS rotational frequency. If 1122 Hz is really the NS rotational frequency, this would give a severe constraint on the EOS.

[8] The empirical relation used in this figure seems to be the relation denoted in [13], for which the prefactor 0.93 becomes 0.96.

2.3.3 The Radius

It is of fundamental astrophysical importance to know the radius of an object such as a NS. Furthermore, the knowledge of the NS radius is important to determine or constrain the NS EOS. It is proved that accurate observation of the *M*–*R* relation or other combination of NS properties enables us to reconstruct the NS EOS [64, 65] by solving the inverse problem. The difficulty is that, however, there exists considerable uncertainties in astronomical observations with electromagnetic radiation. For example, the NS radius is strongly correlated with the distance to the NS, as well as the radii of other astronomical objects do. Emission processes and the atmosphere of the NS are also uncertain, and therefore the systematic error is unavoidably large. By contrast, gravitational-wave astronomy is believed to accomplish the accurate, systematic-error free measurement of the NS radius and/or properties of this kind (say, tidal deformability) in the near future.

The most simple method to determine the NS radius may be to observe the so-called radiation radius, which is defined as

$$R_\infty \equiv R_{\rm NS}\left(1-\frac{2M_{\rm NS}}{R_{\rm NS}}\right)^{-1/2} = R_{\rm NS}(1+z), \tag{2.31}$$

with the gravitational redshift [54],

$$z \equiv \frac{1}{\sqrt{g_{tt}(r=R_{\rm NS})}} - 1 = \left(1-\frac{2M_{\rm NS}}{R_{\rm NS}}\right)^{-1/2} - 1. \tag{2.32}$$

The definition of the radiation radius follows from the Stefan-Boltzmann law of the blackbody radiation. In a flat spacetime, the flux F of a blackbody with the radius $R_{\rm NS}$ and the temperature T is given by

$$F = \left(\frac{R_{\rm NS}}{D}\right)^2 \sigma T^4, \tag{2.33}$$

where σ is the Stefan-Boltzmann constant,[9] for an observer at a distance D. When we take the spacetime curvature into account, the flux and temperature are modified because of the gravitational redshift of the photon and the loss of the photon number flux, which is defined by the number of incoming photons within a unit time interval. Specifically, observed values are given by

$$F_\infty = \frac{F}{(1+z)^2}, \quad T_\infty = \frac{T}{1+z}, \tag{2.34}$$

[9] $\sigma = 5.67 \times 10^{-5}\,{\rm erg\,cm^{-2}\,s^{-1}\,K^{-4}}$.

where it is assumed that the spacetime is flat in the vicinity of the observer. As a result, the "radius" which we can know is the radiation radius defined using a modified Stefan-Boltzmann law by

$$F_\infty = \left(\frac{R_\infty}{D}\right)^2 \sigma T_\infty^4. \tag{2.35}$$

The determination of the radiation radius is often performed by the X-ray observation, and the accuracy suffers from an uncertainty of the atmospheric model even if the distance to the NS is accurately determined by other methods such as parallax measurements. Furthermore, it is found that the optical observation usually gives a larger flux than is expected from the X-ray observation, and this discrepancy is thought to be ascribed to energy redistribution in the NS atmosphere. It should be also noted that the NS radius as a solution of the TOV equation and the photospheric radius of the NS will not agree exactly. Aside from several uncertainties in the radiation radius, simultaneous observation of the radiation radius and the gravitational redshift may give simultaneous estimate of the mass and radius of the NS. Although there exists some observations of the gravitational redshift using absorption lines of atmospheric nuclei [66], it is still too unclear to draw a definite conclusion.

It may be useful to present a recent attempt to estimate the NS radius using the photospheric radius expansion burst in low mass X-ray binaries. The Type-I X-ray burst is a thermonuclear burst of a helium and/or hydrogen layer accreted from the companion, and the photospheric radius expansion burst is a subclass of the Type-I X-ray burst [67]. In a photospheric radius expansion burst, the observed flux, F_∞, first increases, next keeps an approximately constant value, and finally decreases. During the constant peak of the flux, the observed color temperature, T_∞, first decrease and then increase. After the maximum of the color temperature, which is called the "touchdown" moment, the color temperature gradually decreases along with the flux. If we define an emission area by $A \equiv F_\infty/(\sigma T_\infty^4)$, the area first increases and then decreases during the flux peak, and keeps an approximately constant value after the touchdown moment. The method for simultaneous determination of the NS mass, NS radius, and distance to the NS is originally developed in [68], in which the author claimed that the soft equations of state are ruled out using additional information of the gravitational redshift [66]. This model assumes that, in the course of the first decrease of the color temperature, the photosphere expands with the observed Eddington flux,[10] which is seen from an observer at a distance $D \gg R_{\rm NS}$ as

$$F_{\rm edd,\infty} = \frac{M_{\rm NS}}{\kappa_{\rm es} D^2}\sqrt{1 - \frac{2M_{\rm NS}}{r_{\rm ph}}}, \tag{2.36}$$

[10] The observed Eddington flux from z is not reduced by a factor of $1/(1+z)^2$ but by a factor of $1/(1+z)$. Roughly speaking, the difference between the flux and the Eddington flux is a result of stronger "local gravitational force" by a factor of $1+z$ in general relativity.

where $\kappa_{\rm es}$ is the opacity chosen to be that of electron scattering and $r_{\rm ph}$ is a time-varying photospheric radius. Notice that the electron scattering opacity is written as

$$\kappa_{\rm es} = \frac{1+X}{2}\frac{\sigma_{\rm T}}{m_{\rm p}} \approx 0.2(1+X)\,{\rm cm}^2\,{\rm g}^{-1}, \tag{2.37}$$

where X, $\sigma_{\rm T}$, and $m_{\rm p}$ are the hydrogen mass fraction, the Thomson cross section, and the proton mass, respectively.[11] Next, it is assumed that the photosphere contracts to the "touchdown" radius, which is assumed to be the NS radius in [68], keeping the Eddington flux and increasing the color temperature. The values of $F_{\rm edd,\infty}$ and T_∞ are observed at the touchdown moment, and these values are used to estimate the photospheric radius at the touchdown moment using a color correction factor $f_{\rm c}$, which denotes the ratio of the intrinsic color temperature $T_{\rm c}$ to the intrinsic effective temperature $T_{\rm eff}$ as $f_{\rm c} \equiv T_{\rm c}/T_{\rm eff}$. Finally, with additional information such as the gravitational redshift or the distance to the NS, these quantities are combined to give the values of the NS mass and radius. As is described above, [68] used the information of the gravitational redshift, and now it is more common to use an independent measurement of the distance [69].

Whereas this method seems to work, there is several unclear assumptions. The flux is assumed to be the Eddington one, the opacity is assumed to be that of electron scattering and the hydrogen fraction is also assumed, the color correction factor is assumed to be known from a model calculation, and most importantly the photospheric radius at the touchdown moment is identified with the NS radius. It is claimed that the uncertainty of the NS radius is as small as $\lesssim 10\,\%$ within a 1-σ error in [70], but it is also claimed that the uncertainty is as large as $\approx 20\,\%$ within a 1-σ error in [69]. The reason for this is that restrictive assumptions listed above typically lead to imaginary values of $M_{\rm NS}$ and $R_{\rm NS}$ when they are determined by Monte-Carlo simulations. The authors of [70] forced their computation to obtain real values of the mass and radius by restricting parameter spaces, and therefore a relatively tight constraint was obtained. The authors of [69] claimed that these imaginary solutions suggested that the model was internally inconsistent, and obtained looser constraints developing more consistent model by relaxing the assumptions, e.g., $r_{\rm ph} \gg R_{\rm NS}$. If these and other possible unknown systematic uncertainties are well understood, this method will serve as a tool to investigate NS properties such as the NS EOS.

Before closing this section, we describe several bounds on the radius or the compactness of the NS obtained in a purely theoretical manner. First, the radius must be so large that the BH horizon does not appear. This requires

$$R_{\rm NS} > 2M_{\rm NS}, \tag{2.38}$$

assuming that the spin angular momentum of the object is negligible. Second, it is known that the NS radius has to satisfy

---

[11] $\sigma_{\rm T} = 6.65 \times 10^{-25}\,{\rm cm}^2$, $m_{\rm p} = 1.67 \times 10^{-24}$ g.

$$R_{\rm NS} > \frac{9}{4} M_{\rm NS}, \tag{2.39}$$

so that the pressure inside the NS is finite, as long as the energy density decreases to the outward direction [54]. Finally, a more empirical bound on the radius,

$$R_{\rm NS} \gtrsim 2.9 M_{\rm NS}, \tag{2.40}$$

is obtained by requiring that the sound velocity inside the NS does not exceed the speed of light above some matching density, $e_{\rm match}$ [71, 72]. In particular, it is pointed out that all quantities have to scale with the matching density, $e_{\rm match}$, in [72], because this is the only dimensional parameter entering the equation. This finding is consistent with the scaling of the NS maximum mass, (2.28). The top left regions of Fig. 2.2 is excluded by these constraints. As is naturally expected, the *M*–*R* relations computed using nuclear-theory-based EOSs do not enter this prohibited region.

2.4 Piecewise Polytropes

To perform systematic investigations, it is preferable to adopt analytic EOSs rather than tabulated, nuclear-theory-based EOSs, because the tabulated EOSs require the interpolation during numerical computations, and therefore time-consuming. Furthermore, EOSs parametrized by a small number of parameters are preferable from the observational viewpoint, because the early-days observation of gravitational waves will give us only limited information of the NS. For these reasons, several effort has been spent to reproduce important properties of nuclear-theory-based EOSs by analytic EOSs parametrized by a small number of parameters [2, 73, 74].

2.4.1 The Cold-Part Equation of State: Piecewise Polytropes

To model cold, nuclear-theory-based EOSs at high density with a small number of parameters, we employ a piecewise polytropic EOS[12]. This is a phenomenologically parametrized EOS of the form

$$P(\rho) = \kappa_i \rho^{\Gamma_i} \text{ for } \rho_{i-1} \le \rho < \rho_i \ (1 \le i \le n), \tag{2.41}$$

where n is the number of pieces used to parametrize an EOS, ρ_i is the rest-mass density at the boundary of two neighboring ith and $(i+1)$th pieces, κ_i is the polytropic constant for the ith piece, and Γ_i is the adiabatic index for the ith piece. Here,

[12] The piecewise polytropic EOS has also been adopted in some of stellar core-collapse and supernova simulations, but the underlying idea is different. In these problems, the effect of the temperature is not at all negligible, and one-parameter EOSs only include these effects approximately. By contrast, the one-parameter EOS is physically reasonable in the inspiral phase of the binary merger.

$\rho_0 = 0$, $\rho_n \to \infty$, and other parameters $(\rho_i, \kappa_i, \Gamma_i)$ are freely chosen. Requiring the continuity of the pressure at each ρ_i, $2n$ free parameters—say (κ_i, Γ_i)—determine the EOS completely. The specific internal energy, ε, and hence the specific enthalpy, h, are determined by the first law of thermodynamics,

$$\mathrm{d}\varepsilon = \frac{P}{\rho^2}\mathrm{d}\rho, \tag{2.42}$$

$$\mathrm{d}h = \frac{\mathrm{d}P}{\rho}, \tag{2.43}$$

and continuity of each variable at boundary densities, ρ_i. Specifically, ε and h are determined by

$$\varepsilon = \varepsilon_{i-1} + \frac{\kappa_i}{\Gamma_i - 1}\rho^{\Gamma_i - 1}, \tag{2.44}$$

$$h = 1 + \varepsilon_{i-1} + \frac{\kappa_i \Gamma_i}{\Gamma_i - 1}\rho^{\Gamma_i - 1} \tag{2.45}$$

for the ith piece, where the integration constant ε_i is given by

$$\varepsilon_0 = 0, \; \varepsilon_i = \varepsilon_{i-1} + \frac{\kappa_i}{\Gamma_i - 1}\rho_i^{\Gamma_i - 1} - \frac{\kappa_{i+1}}{\Gamma_{i+1} - 1}\rho_i^{\Gamma_{i+1} - 1}. \tag{2.46}$$

It is shown that piecewise polytropic EOSs with four pieces approximately reproduce most properties of the nuclear-theory-based EOSs at high density [2]. Moreover, if we focus on canonical-mass NSs with relatively low central density, the EOS at high density plays a minor role. Thus, we adopt a simplified piecewise polytropic EOS composed of two pieces, one of which models the crust EOS and the other of which the core EOS. This simplification is based on the fact that NSs in the observed binary NSs often have fairly small masses $\lesssim 1.4 M_\odot$ [75] and the maximum rest-mass density in such NSs may not be so high that the EOS at high density plays only a minor role in determining their structure. Furthermore, the maximum rest-mass density inside the NS should only decrease during the evolution of the BH–NS binary due to tidal elongation of the NS by the companion BH.[13]

Table 2.1 lists the EOSs which we employ in this thesis. Following [79] (see also [80–82]), we always fix the EOS for the crust region by parameters below:

$$\Gamma_1 = 1.35692395, \tag{2.47}$$

$$\kappa_1/c^2 = 3.99873692 \times 10^{-8} \left(\mathrm{g\,cm^{-3}}\right)^{1-\Gamma_1}. \tag{2.48}$$

The EOS for the core region is determined by two parameters. One is the adiabatic index of the core EOS, Γ_2. The other parameter is chosen to be the pressure p at a

[13] For the merger of binary NSs, formation of a hypermassive NS with a large central rest-mass density is the frequent outcome [76–78].

Table 2.1 Key ingredients of the adopted EOSs

Model	Γ_2	$\log_{10} p$ (g/cm^3)	$\rho_1 (10^{14}\,\mathrm{g\,cm^{-3}})$	$M_{max}[M_\odot]$	R_{12} (km)	R_{135} (km)	R_{145} (km)
2H	3.0	13.95	0.7033	2.835	15.12	15.23	15.28
1.5H	3.0	13.75	0.9308	2.525	13.63	13.69	13.72
H	3.0	13.55	1.232	2.249	12.25	12.28	12.27
HB	3.0	13.45	1.417	2.122	11.60	11.61	11.59
HBs	2.7	13.45	1.069	1.926	11.67	11.57	11.47
HBss	2.4	13.45	0.6854	1.701	11.74	11.45	11.19
B	3.0	13.35	1.630	2.003	10.98	10.96	10.93
Bs	2.7	13.35	1.269	1.799	10.88	10.74	10.61
Bss	2.4	13.35	0.8547	1.566	10.66	10.27	9.89

Γ_2 is the adiabatic index in the core region and p is the pressure at the fiducial density $\rho_{\rm fidu} = 10^{14.7}\,\mathrm{g\,cm^{-3}}$, which determines the polytropic constant κ_2 of the core region and ρ_1, the critical rest-mass density separating the crust and core regions. M_{max} is the maximum mass of the spherical NS for a given EOS. R_{135}, R_{12}, and R_{145} are the circumferential radius the NS with $M_{NS} = 1.35M_\odot$, $1.2M_\odot$, and $1.45M_\odot$, respectively

fiducial density $\rho_{\rm fidu} = 10^{14.7}\,\mathrm{g\,cm^{-3}}$, because p is closely related to the radius of the NS with a canonical mass [12]. We vary values of these two parameters systematically to investigate the effect of the core EOS. With given values of Γ_2 and p, κ_2 and ρ_1 are determined as

$$\kappa_2 = p\rho_{\rm fidu}^{-\Gamma_2}, \tag{2.49}$$

$$\rho_1 = \left(\frac{\kappa_1}{\kappa_2}\right)^{1/(\Gamma_2-\Gamma_1)}. \tag{2.50}$$

It must be mentioned that the maximum mass constraint, $M_{max} \geq 1.97 \pm 0.04M_\odot$, reported by [53] is not satisfied for the EOS with $\Gamma_2 \neq 3$. In spite of this deficit,[14] the parameter sets in Table 2.1 are still meaningful to investigate dependence of gravitational waves and merger remnants on the EOS.

Figure 2.3 shows the M–R relation, i.e., the relation between the mass, M_{NS}, and circumferential radius, R_{NS}, of spherical NSs for piecewise polytropic EOSs adopted in this study, as well as for a $\Gamma = 2$ polytropic EOS with $\kappa/c^2 = 2 \times 10^{-16}\,\mathrm{g^{-1}\,cm^3}$. Figure 2.3 shows that for a given mass $\sim 1.35M_\odot$, the radius depends strongly on the EOSs, whereas the radius for a given piecewise polytropic EOS depends only weakly on the mass around the canonical mass $\sim 1.35M_\odot$. This weak dependence of the radius on the mass is an often-seen feature for the nuclear-theory-based EOSs [13]. By contrast, the relation calculated with the $\Gamma = 2$ polytropic EOS does not show this feature. Figure 2.3 illustrates that the dependence of the radius, R_{NS}, on the mass, M_{NS}, is much stronger for the $\Gamma = 2$ polytropic EOS than for the piecewise polytropic EOSs. This illustrates that the $\Gamma = 2$ polytropic EOS is not very realistic.

[14] When we started this study, the NS mass of J1614-2230 was not known.

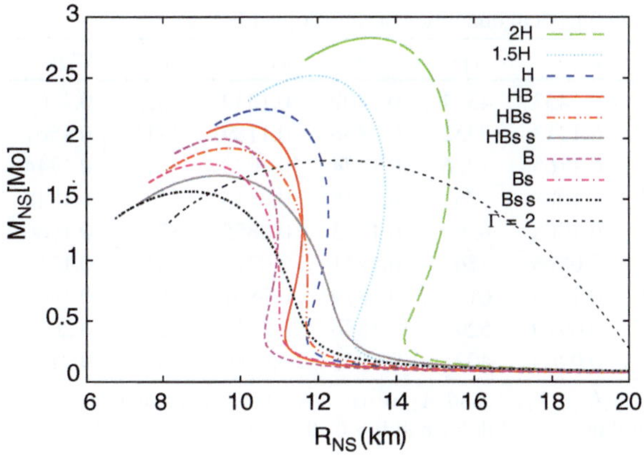

Fig. 2.3 The relation between the mass and circumferential radius of spherical NSs for the piecewise polytropic EOSs adopted in this study. For comparison, the relation for a $\Gamma = 2$ polytropic EOS with $\kappa/c^2 = 2 \times 10^{-16}\,\mathrm{g}^{-1}\,\mathrm{cm}^3$ is also shown

Comparison of the quantities among HB, HBs, and HBss EOS models in Table 2.1 reveals a complicated *M–R* relation: HB is not always stiffer[15] than HBss. Indeed, the radius with $M_{\rm NS} = 1.2M_\odot$ is largest for HBss and smallest for HB among three models, whereas the radius with $M_{\rm NS} = 1.35M_\odot$ is largest for HB and smallest for HBss. This complicated relation of the "stiffness" is due to the choice for the combination (Γ_2, p) (cf. Table 2.1). For a density smaller than $\boldsymbol{\rho}_{\rm fidu}$, HBss EOS is stiffer than HB and HBs EOSs, whereas for a high density $\boldsymbol{\rho} > \boldsymbol{\rho}_{\rm fidu}$, HB EOS is stiffer than the others. For a given high-mass NS for which the central density is much larger than $\boldsymbol{\rho}_{\rm fidu}$, the radius with HB EOS should be larger than that with other two EOSs. By contrast, for a given low-mass NS for which the central density is not very high, the radius with HB EOS should be the smallest.

Table 2.2 lists the compactness, $l = 2$ tidal Love number, and $l = 2$ nondimensional tidal deformability for the EOS adopted in this study. Notice that the tidal deformability, λ, is related to Λ by

$$\lambda \simeq 4.72 \times 10^{36} \left(\frac{\Lambda}{1000}\right) \left(\frac{M_{\rm NS}}{1.35M_\odot}\right)^5 \mathrm{g\,cm^2\,s^2}, \tag{2.51}$$

which results from (2.12). By comparing $1.35M_\odot$ NSs for HB, HBs, and HBss EOSs, it is found that the tidal Love number depends strongly on the EOS, especially on Γ_2, even for NSs with similar values of the compactness. The fact that a larger value of Γ_2 leads to a larger value of k_2 indicates that the NS is more susceptible to the tidal force when the adiabatic index of the EOS is larger [83]. By contrast, it is also found that the tidal Love number depends only weakly on the value of p for NSs with

[15] In this thesis, the stiffness is simply determined by the magnitude of pressure for the nuclear-density region. We do not determine it by the adiabatic index.

Table 2.2 Important physical quantities of the adopted EOSs

Model	$\mathscr{C}_{12}$	k_{12}	Λ_{12}	$\mathscr{C}_{135}$	k_{135}	Λ_{135}	$\mathscr{C}_{145}$	k_{145}	Λ_{145}
2H	0.1172	0.1453	4384	0.1309	0.1342	2325	0.1401	0.1265	1560
1.5H	0.1301	0.1313	2352	0.1456	0.1189	1211	0.1561	0.1105	795
H	0.1447	0.1165	1226	0.1624	0.1029	607	0.1744	0.0937	387
HB	0.1527	0.1088	873	0.1718	0.0946	422	0.1848	0.0851	263
HBs	0.1519	0.1005	829	0.1723	0.0855	375	0.1866	0.0754	222
HBss	0.1509	0.0886	756	0.1741	0.0723	301	0.1914	0.0610	158
B	0.1614	0.1010	615	0.1819	0.0861	289	0.1960	0.0763	176
Bs	0.1629	0.0910	529	0.1856	0.0752	228	0.2017	0.0645	129
Bss	0.1663	0.0765	402	0.1940	0.0586	142	0.2164	0.0456	64

$\mathscr{C}_{135}(\mathscr{C}_{12}, \mathscr{C}_{145})$, $k_{135}(k_{12}, k_{145})$, and $\Lambda_{135}(\Lambda_{12}, \Lambda_{145})$ are the compactness, $l = 2$ tidal Love number, and $l = 2$ nondimensional deformability of the NS with $M_{\rm NS} = 1.35M_\odot(1.2M_\odot, 1.45M_\odot)$, respectively

similar values of compactness. This fact provokes the dependence of gravitational waves on the values of both p and Γ_2, not merely on the NS compactness.

2.4.2 Thermal Corrections

In dynamical simulations, the matter inside the NS becomes hot due to shock heating, especially after the tidal disruption. Typically, the average temperature of the accretion disk formed after the BH–NS merger becomes as high as several MeV [84, 85]. In such a phase, cold EOSs are no longer sufficient to model the material, and we have to take the thermal effect into account. Because we are mainly focus on gravitational waves and on prompt formation of accretion disks in this study, implementation of a nuclear-theory-based finite-temperature EOS [77, 78] is beyond the scope of this thesis. Therefore, we adopt a simple Γ-law, ideal-gas EOS for the thermal part as a correction to the dominant, zero-temperature part. We first decompose the pressure and specific internal energy into cold and thermal parts as

$$P = P_{\rm cold} + P_{\rm th}, \ \varepsilon = \varepsilon_{\rm cold} + \varepsilon_{\rm th}. \tag{2.52}$$

We calculate the cold parts of both variables using the piecewise polytropic EOS from the baryon rest-mass density, ρ, and then the thermal part of the specific internal energy is defined from ε as

$$\varepsilon_{\rm th} \equiv \varepsilon - \varepsilon_{\rm cold}. \tag{2.53}$$

Because $\varepsilon_{\rm th}$ vanishes in the absence of shock heating, $\varepsilon_{\rm th}$ is regarded as the finite-temperature part. Finally, we compute the thermal part of the pressure according to

$$P_{\rm th} = (\Gamma_{\rm th} - 1)\rho\varepsilon_{\rm th}, \tag{2.54}$$

where we choose $\Gamma_{\rm th}$ equal to the adiabatic index in the crust region, Γ_1, for simplicity. The effect of different choices of $\Gamma_{\rm th}$ should be investigated in the near future [86].

We can approximately estimate the temperature T of material as

$$\varepsilon_{\rm th} c^2 = \frac{3kT}{2m_{\rm u}} + \frac{11 a_{\rm r} T^4}{4\rho}, \tag{2.55}$$

where we inserted c for clarity, by assuming that the pressure is composed of the gas pressure of free nuclei and the radiation pressure of photons, relativistic electrons, and relativistic positrons [87]. Here, $m_{\rm u}$ and $a_{\rm r}$ are the atomic mass unit and the radiation-density constant, respectively.[16] It should be noted that the gas and radiation pressure becomes equal at the temperature

$$T_{\rm c} \equiv \left(\frac{6k\rho}{11 a_{\rm r} m_{\rm u}}\right)^{1/3} \simeq 4 \times 10^{10} \left(\frac{\rho}{10^{10}\,{\rm g\,cm^{-3}}}\right)^{1/3} \ {\rm K}, \tag{2.56}$$

and the gas pressure is dominant below this critical temperature. Figure 2.4 shows this relation for the rest-mass density $\rho = 10^{10}$, 10^{11}, and $10^{12}\,{\rm g\,cm^{-3}}$. More accurate expression is obtained by considering the constituent nuclei such as α-particle for gas pressure, nonzero degeneracies of electrons and positrons for radiation pressure, and optical depths of neutrinos, which may be trapped and contribute to the pressure when $T \gtrsim 10^{11}$ K and $\rho \gtrsim 10^{12}\,{\rm g\,cm^{-3}}$, for radiation pressure [88, 89]. Especially,

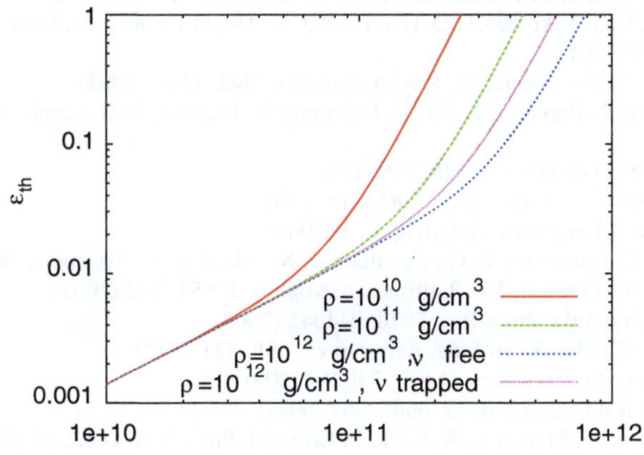

Fig. 2.4 An approximate relation (2.55) between the temperature T and the thermal part of the specific internal energy $\varepsilon_{\rm th}$. The curves for $\rho = 10^{10}$, 10^{11}, and $10^{12}\,{\rm g\,cm^{-3}}$ are shown. The *curve* for $\rho = 10^{12}\,{\rm g\,cm^{-3}}$ with trapped neutrinos and antineutrinos is also shown. Notice that $T = 10^{10}\,{\rm K} \approx 1\,{\rm MeV}$, therefore electrons and positrons are safely assumed to be relativistic in the temperature range shown here

[16] $m_{\rm u} = 1.66 \times 10^{-24}\,{\rm g}$, $a_{\rm r} = 7.57 \times 10^{-15}\,{\rm erg\,cm^{-3}\,K^{-4}}$.

the prefactor $11/4 = 1 + 2 \times (7/8)$ of the radiation pressure becomes $1 + 2 \times (7/8) + 6 \times (7/8) = 8$ if neutrinos and antineutrinos of all the flavors are trapped. Figure 2.4 also includes the curve for $\rho = 10^{12}\,\mathrm{g\,cm^{-3}}$ with trapped neutrinos and antineutrinos.

References

1. S.L. Shapiro, S.A. Teukolsky, *Black holes, white dwarfs, and neutron stars: the physics of compact objects* (wiley-vch, 1983)
2. J.S. Read, B.D. Lackey, B.J. Owen, J.L. Friedman, Phys. Rev. D **79**, 124032 (2009)
3. K. Sato, Prog. Theor. Phys. **54**, 1325 (1975)
4. A. Burrows, J.M. Lattimer, Astrophys. J. **307**, 178 (1986)
5. J.A. Pons, S. Reddy, M. Prakash, J.M. Lattimer, J.A. Miralles, Astrophys. J. **513**, 780 (1999)
6. A. Mezzacappa, A.C. Calder, S.W. Bruenn, J.M. Blondin, M.W. Guidry, M.R. Strayer, A.S. Umar, Astrophys. J. **493**, 848 (1998)
7. L. Dessart, A. Burrows, E. Livne, C.D. Ott, Astrophys. J. **645**, 534 (2006)
8. D.G. Yakovlev, C.J. Pethick, Annu. Rev. Astron. Astrophys. **42**, 169 (2004)
9. D. Page, J.M. Lattimer, M. Prakash, A.W. Steiner, Astrophys. J. Suppl. **155**, 623 (2004)
10. J.M. Lattimer, K.A. van Riper, M. Prakash, M. Prakash, Astrophys. J. **425**, 802 (1994)
11. J.M. Lattimer, C.J. Pethick, M. Prakash, P. Haensel, Phys. Rev. D **66**, 21 (1991)
12. J.M. Lattimer, M. Prakash, Astrophys. J. **550**, 426 (2001)
13. J.M. Lattimer, M. Parakash, Phys. Rep. **442**, 109 (2007)
14. J.M. Lattimer, M. Prakash, Science **304**, 536 (2004)
15. C.O. Heinke, W.C.G. Ho, Astrophys. J. **719**, L167 (2010)
16. D. Page, M. Prakash, J.M. Lattimer, A.W. Steiner, Phys. Rev. Lett. **106**, 081101 (2011)
17. P.S. Shternin, D.G. Yakovlev, C.O. Heinke, W.C.G. Ho, D.J. Patnaude, Mon. Not. Roy. Astron. Soc. **412**, L108 (2011)
18. R. Voss, T.M. Tauris, Mon. Not. Roy. Astron. Soc. **342**, 1169 (2003)
19. K. Belczynski, R. Perna, T. Bulik, V. Kalogera, N. Ivanova, D.Q. Lamb, Astrophys. J. **648**, 1110 (2006)
20. K.A. van Riper, Astrophys. J. **329**, 339 (1988)
21. D. Lai, S.L. Shapiro, Astrophys. J. **383**, 745 (1991)
22. N. Stergioulas, Living. Rev. Relativity **6**, 3 (2003)
23. M. Bocquet, S. Bonazzola, E. Gourgoulhon, J. Novak, Astron. Astrophys. **301**, 757 (1995)
24. C.Y. Cardall, M. Prakash, J.M. Lattimer, Astrophys. J. **554**, 322 (2001)
25. K. Kiuchi, S. Yoshida, Phys. Rev. D **78**, 044045 (2008)
26. K. Kiuchi, K. Kotake, S. Yoshida, Astrophys. J **698**, 541 (2009)
27. K. Ioka, M. Sasaki, Phys. Rev. D **67**, 124026 (2003)
28. K. Ioka, M. Sasaki, Astrophys. J. **600**, 296 (2004)
29. E. Gourgoulhon, C. Markakis, K. Uryū, Y. Eriguchi, Phys. Rev. D **83**, 104007 (2011)
30. R.C. Tolman, Phys. Rev. **55**, 364 (1939)
31. J.R. Oppenheimer, G.M. Volkoff, Phys. Rev. **55**, 374 (1939)
32. R.F. Tooper, Astrophys. J. **140**, 434 (1964)
33. H.A. Buchdahl, Astrophys. J. **140**, 1512 (1964)
34. H.A. Buchdahl, Astrophys. J. **147**, 310 (1967)
35. É.É. Flanagan, T. Hinderer, Phys. Rev. D **77**, 021502(R) (2008).
36. D. Lai, F.A. Rasio, S.L. Shapiro, Astrophys. J. **406**, L63 (1993)
37. D. Lai, F.A. Rasio, S.L. Shapiro, Astrophys. J. **420**, 811 (1994)
38. B.D. Lackey, K. Kyutoku, M. Shibata, P.R. Brady, J.L. Friedman, Phys. Rev. D **85**, 044061 (2012)

39. R.A. Brooker, T.W. Olle, Mon. Not. Roy. Astron. Soc. **115**, 101 (1955)
40. Z. Kopal, *Close binary systems* (Wiley, New York, 1959)
41. T. Binnington, E. Poisson, Phys. Rev. D **80**, 084018 (2009)
42. T. Damour, A. Nagar, Phys. Rev. D **80**, 084035 (2009)
43. T. Hinderer, Astrophys. J. **677**, 1216 (2008)
44. T. Hinderer, Astrophys. J. **697**, 964 (2009)
45. T. Regge, J.A. Wheeler, Phys. Rev. **108**, 1063 (1957)
46. K.S. Thorne, A. Campolattaro, Astrophys. J. **149**, 591 (1967)
47. S. Postnikov, M. Prakash, J.M. Lattimer, Phys. Rev. D **82**, 024016 (2010)
48. K.S. Thorne, Rev. Mod. Phys. **52**, 299 (1980)
49. K.S. Thorne, J.B. Hartle, Phys. Rev. D **31**, 1815 (1985)
50. C.E. Rhoades, R. Ruffini, Phys. Rev. Lett. **32**, 324 (1974)
51. J.B. Hartle, Phys. Rep. **46**, 201 (1978)
52. J.B. Hartle, A.G. Sabbadini, Astrophys. J. **213**, 831 (1977)
53. P. Demorest, T. Pennucci, S. Ransom, M. Roberts, J. Hessels, Nature **467**, 1081 (2010)
54. R.M. Wald, *General Relativity* (The university of Chicago press, 1984).
55. J.W.T. Hessels, S.M. Ransom, I.H. Stairs, P.C.C. Freire, V.M. Kaspi, F. Camilo, Science **311**, 1901 (2006)
56. M.H. van Kerkwijk, R.P. Breton, S.R. Kulkarni, Astrophys. J. **728**, 95 (2011)
57. I.I. Shapiro, Phys. Rev. Lett. **13**, 789 (1964)
58. J.L. Friedman, J.R. Ipser, L. Parker, Astrophys. J. **304**, 115 (1986)
59. J.L. Friedman, J.R. Ipser, L. Parker, Astrophys. J. **351**, 705 (1990)
60. P. Haensel, J.L. Zdunik, Nature **340**, 617 (1989)
61. P. Haensel, M. Salgado, S. Bonazzola, Astron. Astrophys. **296**, 745 (1985)
62. P. Haensel, J.L. Zdunik, M. Bejger, J.M. Lattimer, Astron. Astrophys. **502**, 605 (2009)
63. P. Kaaret, Z. Prieskorn, J.J.M. in 't Zand, S. Brandt, N. Lund, S. Mereghetti, D. Götz, E. Kuulkers, J.A. Tomsick. Astrophys. J. **657**, L97 (2007)
64. L. Lindblom, Astrophys. J. **398**, 569 (1992)
65. T. Harada, Phys. Rev. C **64**, 048801 (2001)
66. J. Cottam, F. Paerels, M. Mendez, Nature **420**, 51 (2002)
67. T. Strohmayer, L. Bildsten, in Compact Stellar X-ray Sources, ed. by W.H.G. Lewin, M. van der Klis (Cambridge University Press, 2006)
68. F. Özel, Nature **441**, 1115 (2006)
69. A.W. Steiner, J.M. Lattimer, E.F. Brown, Astrophys. J. **722**, 33 (2010)
70. F. Özel, G. Baym, T. Güver, Phys. Rev. D 82, 101301(R) (2010).
71. L. Lindblom, Astrophys. J. **278**, 364 (1984)
72. S. Koranda, N. Stergioulas, J.L. Friedman, Astrophys. J. **488**, 799 (1997)
73. F. Özel, D. Psaltis, Phys. Rev. D **80**, 103003 (2009)
74. L. Lindblom, Phys. Rev. D **82**, 103011 (2010)
75. I.H. Stairs, Science **304**, 547 (2004)
76. K. Hotokezaka, K. Kyutoku, H. Okawa, M. Shibata, K. Kiuchi, Phys. Rev. D **83**, 124008 (2011)
77. Y. Sekiguchi, K. Kiuchi, K. Kyutoku, M. Shibata, Phys. Rev. Lett. **107**, 051102 (2011)
78. Y. Sekiguchi, K. Kiuchi, K. Kyutoku, M. Shibata, Phys. Rev. Lett. **107**, 211101 (2011)
79. J.S. Read, C. Markakis, M. Shibata, K. Uryū, J.D.E. Creighton, J.L. Friedman, Phys. Rev. D **79**, 124033 (2009)
80. K. Kyutoku, M. Shibata, K. Taniguchi, Phys. Rev. D **82**, 044049 (2010)
81. K. Kyutoku, M. Shibata, K. Taniguchi, Phys. Rev. D **84**, 049902(E) (2011).
82. K. Kyutoku, H. Okawa, M. Shibata, K. Taniguchi, Phys. Rev. D **84**, 064018 (2011)
83. M. Ishii, M. Shibata, Y. Mino, Phys. Rev. D **71**, 044017 (2005)
84. M. Shibata, K. Taniguchi, Phys. Rev. D **73**, 064027 (2006)
85. F. Foucart, M.D. Duez, L.E. Kidder, S.A. Teukolsky, Phys. Rev. D **83**, 024005 (2011)
86. A. Bauswein, H.T. Janka, R. Oechslin, Phys. Rev. D **82**, 084043 (2010)
87. M. Shibata, K. Taniguchi, Phys. Rev. D **77**, 084015 (2008)
88. W.H. Lee, E. Ramirez-Ruiz, D. Page, Astrophys. J. **632**, 421 (2005)
89. M. Shibata, Y.I. Sekiguchi, R. Takahashi, Prog. Theor. Phys. **118**, 257 (2007)

Chapter 3
Computing Initial Conditions

We employ BH–NS binaries in quasiequilibrium states for initial conditions of our numerical simulations. In this chapter, we describe the formulation for the computation of a quasiequilibrium state. The details of the formulation and numerical methods, except for the issues on the BH spin, are described in [1], and the issues related to the BH spin are described in [2]. Computations of the quasiequilibrium states are performed using the spectral-method library `LORENE` [3].

3.1 Assumption

We compute a quasiequilibrium state of the BH–NS binary as a solution of the initial value problem of general relativity [4]. As far as the orbital separation d is large enough, the time scale $t_{\rm GW}$ of the orbital contraction due to the gravitational radiation reaction is much longer than the orbital period $P_{\rm orb}$. Specifically, $t_{\rm GW}$ is given by a lifetime of the binary, (1.10), and $P_{\rm orb}$ is given by (1.2) in Newtonian gravity. Hence, the ratio of $t_{\rm GW}$ to $P_{\rm orb}$ is written as

$$\frac{t_{\rm GW}}{P_{\rm orb}} \simeq 1.1 \left(\frac{d}{6m_0}\right)^{5/2} \left(\frac{(1+Q)^2}{4Q}\right). \tag{3.1}$$

In numerical simulations of the binary coalescences, we have to track $\gtrsim$5 orbits in order to calculate accurate gravitational waveforms during the late inspiral and merger phases, and hence, the orbital separation of the initial configuration has to be so large that we may safely neglect the gravitational radiation reaction. Thus, we give a BH–NS binary in a quasicircular orbit, i.e., the binary in an approximate equilibrium state in the corotating frame. To satisfy the quasiequilibrium requirements described above, we assume the existence of a helical Killing vector with the orbital angular velocity Ω,

$$\xi^\mu = (\partial_t)^\mu + \Omega(\partial_\varphi)^\mu, \tag{3.2}$$

where ∂_φ is a generator of the rotation.

K. Kyutoku, *The Black Hole–Neutron Star Binary Merger in Full General Relativity*, Springer Theses, DOI: 10.1007/978-4-431-54201-8_3, © Springer Japan 2013

We also assume that the NS is in the hydrostatic equilibrium in the corotating frame, and has an irrotational velocity field, which is believed to be a reliable approximation to an astrophysically realistic configuration of a compact binary just before the merger. One of the reason for this is that the time scale of the NS spin-up due to the tidal synchronization (tidal locking) is longer than the time scale of the orbital contraction due to the gravitational radiation reaction, as far as the viscosity in the NS is not extremely large [5, 6]. The other reason is that the typical observed rotational period of the NS, $P_{\rm rot} \gtrsim 1$ s, is much longer than the orbital period of the quasi-equilibrium binary just before the merger. Furthermore, the NS rotational period is assumed to increase secularly due to the magnetic dipole radiation [7], and formation of a millisecond pulsar is hardly expected for the BH–NS binary due to the absence of a recycling process, which may be driven by the accretion of material from a less-evolved companion.[1]

3.2 The Initial Value Problem of General Relativity

Because the Einstein equations, (1.32), are written in a fully covariant manner, what "solving the Einstein equations" means is highly nontrivial in general situations. One attractive approach is to reformulate the Einstein equations from the perspective of a Cauchy problem, in which given initial data evolves in time. This approach is called the 3+1 formalism of general relativity [12], on which most formulations of numerical relativity rely (see [13–15] for reviews). In this section, we review the 3+1 formalism and initial value problem of general relativity, and present the formulation to solve the initial value problem.

3.2.1 The 3+1 Formalism

In the 3+1 formalism, the spacetime is considered to be foliated by a one-parameter family of spatial hypersurfaces Σ_t, or a foliation $\{\Sigma_t\}$. In this formalism, we first prepare initial data on a given initial hypersurface Σ_0, and perform dynamical simulations to evolve the data into the future. One of fundamental geometric objects in the 3+1 formalism is a timelike unit normal vector n^μ to each slice, Σ_t, and an observer whose four velocity is n^μ is called the Eulerian observer. The direction of the coordinate time is denoted by a time vector t^μ, which is written as

$$t^\mu = (\partial_t)^\mu = \alpha n^\mu + \beta^\mu \, , \; \beta^\mu n_\mu = 0, \tag{3.3}$$

where α and β^μ are called the lapse function and the shift vector, respectively. The lapse function and the shift vector represent the coordinate or gauge freedom to

[1] The recycling may be possible for binary NSs. Some effort to construct quasiequilibrium states with arbitrary circulation is devoted [8–11].

choose the time and spatial coordinates, respectively. The metric on Σ_t induced by $g_{\mu\nu}$, which is called the induced metric (or the first fundamental form), is given by

$$\gamma_{\mu\nu} \equiv g_{\mu\nu} + n_\mu n_\nu. \tag{3.4}$$

Using these variables, the line element in the spacetime is expressed as

$$\mathrm{d}s^2 = g_{\mu\nu}\mathrm{d}x^\mu \mathrm{d}x^\nu = -\alpha^2 \mathrm{d}t^2 + \gamma_{ij}(\mathrm{d}x^i + \beta^i \mathrm{d}t)\,(\mathrm{d}x^j + \beta^j \mathrm{d}t). \tag{3.5}$$

Now, solving the Einstein equations is reformulated to computing the time evolution of γ_{ij}, where the gauge variables (α, β^i) are chosen freely. The complete description of the foliation requires the extrinsic curvature (or the second fundamental form), which is defined by

$$K_{\mu\nu} \equiv -\frac{1}{2}\pounds_n \gamma_{\mu\nu}. \tag{3.6}$$

Roughly speaking, γ_{ij} and K_{ij} denote the "position" and "velocity" of the geometry, respectively.[2] In fact, the Einstein equations give evolution equations of K_{ij}, because the Einstein equations are second-order differential equations. Decomposing the energy-momentum tensor with respect to the Eulerian observer as

$$\rho_{\mathrm{H}} \equiv T^{\alpha\beta} n_\alpha n_\beta = \rho h(\alpha u^t)^2 - P, \tag{3.7}$$

$$j_\mu \equiv -\gamma_{\mu\alpha} T^{\alpha\beta} n_\beta = \rho h(\alpha u^t)(\gamma_{\mu\alpha} u^\alpha), \tag{3.8}$$

$$S_{\mu\nu} \equiv \gamma_{\mu\alpha}\gamma_{\nu\beta} T^{\alpha\beta} = \rho h(\gamma_{\mu\alpha} u^\alpha)(\gamma_{\nu\beta} u^\beta) + P\gamma_{\mu\nu}, \tag{3.9}$$

we obtain the Hamiltonian constraint,

$$R + K^2 - K_{ij}K^{ij} = 16\pi\rho_{\mathrm{H}}, \tag{3.10}$$

the momentum constraint,

$$D_j(K^{ij} - K\gamma^{ij}) = 8\pi j^i, \tag{3.11}$$

the evolution equation of γ_{ij},

$$\partial_t \gamma_{ij} = -2\alpha K_{ij} + \pounds_\beta \gamma_{ij}, \tag{3.12}$$

and the evolution equation of K_{ij},

$$\partial_t K_{ij} = -D_i D_j \alpha + \alpha\left[R_{ij} - 2K_{ik}K_j{}^k + KK_{ij} - 8\pi\left(S_{ij} - \frac{1}{2}S\gamma_{ij}\right) - 4\pi\rho_{\mathrm{H}}\gamma_{ij}\right] + \pounds_\beta K_{ij}. \tag{3.13}$$

---

[2] This viewpoint becomes clear in the canonical formulation of general relativity [12] (see the Appendix A).

The Hamiltonian and momentum constraints, (3.10) and (3.11), do not involve second-order time derivatives of γ_{ij}, and this is the very reason why they are called the constraint equations. The evolution equation of γ_{ij}, (3.12), denotes the kinematic relation between the "position" and "velocity," and the evolution equation of K_{ij}, (3.13), is the dynamical equation obtained from the Einstein equations.

3.2.2 Solving the Initial Value Problem

Initial data of the metric consists of (γ_{ij}, K_{ij}) on a given initial hypersurface. For these quantities to represent physical spacetimes, they must satisfy the Einstein constraint equations. In other words, we have to solve the Hamiltonian and momentum constraints to give the initial data, (γ_{ij}, K_{ij}), on an initial hypersurface with appropriate physical settings. Here, the subtlety arises from the consideration of the number of degrees of freedom (d.o.f). Whereas we have only four constraint equations, we have to give twelve components of (γ_{ij}, K_{ij}). Therefore, it is required to decompose the eight freely chosen d.o.f and the four constrained d.o.f. to solve the constraints. This problem is called the initial value problem of general relativity (see [4] and reference therein for reviews). Furthermore, it is desirable for initial data of a binary to satisfy quasiequilibrium conditions in a corotating frame. For this purpose, we want to determine initial data of (α, β^i) employing some parts of the Einstein evolution equations. In this study, we compute the induced metric γ_{ij}, the extrinsic curvature K_{ij}, the lapse function α, and the shift vector β^i by a mixture of the extended conformal thin-sandwich approach (XTCS) [16, 17] and the conformal transverse-traceless (CTT) decomposition [13, 18, 19] in the puncture framework [20–22].

In the CTT approach, the induced metric is decomposed into a conformal factor ψ and a background metric $\hat{\gamma}_{ij}$ as

$$\gamma_{ij} = \psi^4 \hat{\gamma}_{ij}. \tag{3.14}$$

We denote the covariant derivative associated with $\hat{\gamma}_{ij}$ as $\hat{D}_i$. The extrinsic curvature is also decomposed into trace, vector, and transverse-traceless tensor parts as

$$\begin{aligned} K^{ij} &= A^{ij} + \frac{1}{3} K \gamma^{ij} \\ &= \psi^{-10} \hat{A}^{ij} + \frac{1}{3} K \gamma^{ij} \\ &= \psi^{-10} \left[\hat{A}^{ij}_{\rm TT} + \left(\hat{\mathbb{L}} W \right)^{ij} \right] + \frac{1}{3} K \gamma^{ij}, \end{aligned} \tag{3.15}$$

where $\hat{\mathbb{L}}$ is the longitudinal derivative operator defined by

$$\left(\hat{\mathbb{L}} W \right)^{ij} \equiv \hat{D}^i W^j + \hat{D}^j W^i - \frac{2}{3} \hat{\gamma}^{ij} \hat{D}_k W^k, \tag{3.16}$$

and the symbol "TT" stands for transverse-traceless, i.e.,

$$\hat{\gamma}^{ij}\hat{A}^{ij}_{\mathrm{TT}} = 0\,,\ \hat{D}_j\hat{A}^{ij}_{\mathrm{TT}} = 0. \tag{3.17}$$

It should be noted here that the index of W^i is raised and lowered by the background metric. The weight ψ^{-10} is introduced to preserve transverse-traceless nature of $\hat{A}^{ij}_{\mathrm{TT}}$ before and after the conformal transformation. Now we give $(\hat{\gamma}_{ij}, K, \hat{A}^{ij}_{\mathrm{TT}})$ as freely specifiable data, and determine (ψ, W^i) from the Einstein constraint equations. According to the decomposition above, the constraint equations, (3.10) and (3.11), are rewritten to give

$$\hat{D}^2\psi = \frac{1}{8}\psi\hat{R} + \frac{1}{12}\psi^5K^2 - \frac{1}{8}\psi^{-7}\hat{A}_{ij}\hat{A}^{ij} - 2\pi\psi^5\rho_{\mathrm{H}}, \tag{3.18}$$

$$\hat{\Delta}_{\mathbb{L}}W^i = \frac{2}{3}\psi^6D^iK + 8\pi\psi^{10}j^i, \tag{3.19}$$

where $\hat{R}$ is the scalar curvature of $\hat{\gamma}_{ij}$ and $\hat{\Delta}_{\mathbb{L}}$ denotes the vectorial Laplacian operator defined by

$$\hat{\Delta}_{\mathbb{L}}W^i \equiv D_j\left(\hat{\mathbb{L}}W\right)^{ij} = \hat{D}^2W^i + \frac{1}{3}\hat{D}^i\hat{D}_jW^j + \hat{R}_j{}^iW^j. \tag{3.20}$$

Equation (3.18) is called the York-Lichnerowicz equation. Among the freely specifiable data, the maximal slicing condition $K = 0$ and an approximate no-radiation condition $\hat{A}^{ij}_{\mathrm{TT}} = 0$ are often adopted to compute initial data. The choice of $\hat{\gamma}_{ij}$ is a matter of debate for a long time, and the conformal flatness condition $\hat{\gamma}_{ij} = f_{ij}$ is often adopted by virtue of its simplicity and expectation that this condition also corresponds to an approximate no-radiation condition.

The XCTS approach[3] allows us to impose quasiequilibrium conditions on the background metric, $\hat{\gamma}_{ij}$, and the trace of the extrinsic curvature, K. This is accomplished by specifying the time evolution of these quantities, which is not restricted to the quasiequilibrium conditions in general, on the initial hypersurface. Because the background metric has only five d.o.f due to the Hamiltonian constraint, we specify the time derivative of $\hat{\gamma}_{ij}$ up to this d.o.f by imposing traceless condition as

$$\hat{u}_{ij} \equiv \partial_t\hat{\gamma}_{ij}\,,\ \hat{\gamma}^{ij}\hat{u}_{ij} = 0. \tag{3.21}$$

This restriction automatically ensures stationarity of the determinant, $\partial_t\hat{\gamma} = 0$. Now freely specifiable data are $(\hat{\gamma}_{ij}, \hat{u}_{ij}, K)$. The conformal factor, ψ, is again determined by solving the Hamiltonian constraint. The momentum constraint is rewritten as an elliptic equation to determine the shift vector as

---

[3] In the original CTS approach proposed in [16], only the quasiequilibrium condition on $\hat{\gamma}_{ij}$ is imposed. Later, this formalism is extended to give the equilibrium condition on K in [17]. The XCTS essentially shares the formulation of historic papers, [23, 24].

$$\hat{\Delta}_{\mathbb{L}}\beta^i - \left(\hat{\mathbb{L}}\beta\right)^{ij}\hat{D}_j\left(\alpha\psi^{-6}\right) = \frac{4}{3}\alpha\hat{D}^i K + \alpha\psi^{-6}\hat{D}_j\left(\frac{\psi^6}{\alpha}\hat{u}^{ij}\right) + 16\pi\alpha\psi^4 j^i, \tag{3.22}$$

using the fact that the conformally-weighted trace-part of the extrinsic curvature is written as

$$\hat{A}^{ij} = \frac{\psi^6}{2\alpha}\left[\left(\hat{\mathbb{L}}\beta\right)^{ij} - \hat{u}^{ij}\right]. \tag{3.23}$$

Notice that in the standard (X)CTS approach, it is not required to decompose $\hat{A}_{ij}$ into vector and tensor parts. In order to determine the lapse function, α, time evolution of the trace of the extrinsic curvature, $\partial_t K$, is specified. The evolution equation of K is derived by (3.12) and (3.13), and it turns out to give an elliptic equation to determine α when $\partial_t K$ is specified. Using the Hamiltonian constraint and performing the conformal transformation, we obtain

$$\begin{aligned}\hat{D}^2(\alpha\psi) &= \frac{1}{8}\alpha\psi\hat{R} + \frac{5}{12}\alpha\psi^{-3}K^2 + \frac{7}{8}\alpha\psi^{-7}\hat{A}_{ij}\hat{A}^{ij} \\ &\quad - \psi^5(\partial_t - £_\beta)K + 2\pi\alpha\psi^5(\rho_{\rm H} + 2S).\end{aligned} \tag{3.24}$$

Here, we write this equation as the elliptic equation to determine $\alpha\psi$, rather than α itself, because source terms for $\alpha\psi$ falls off more rapidly at spatial infinity than for α. In the following of this section, we always assume the conformal flatness of the induced metric,[4] the maximal slicing condition, and their preservation in time as follows:

$$\hat{\gamma}_{ij} = f_{ij}\,,\ K = 0\,,\ \hat{u}_{ij} = 0\,,\ \partial_t K = 0. \tag{3.25}$$

These quasiequilibrium conditions have to be imposed along the direction of the helical symmetry as $£_\xi\hat{\gamma}_{ij} = 0$ and $£_\xi K = 0$, and they can be replaced by $\partial_t\hat{\gamma}_{ij} = 0$ and $\partial_t K = 0$ in the conformal flatness approximation.

3.2.3 The Puncture Framework

To avoid divergence associated with the BH in the numerical computation, we adopt the puncture method [20, 26, 27]. The conformal factor, ψ, and a weighted lapse function $\Phi \equiv \alpha\psi$ is decomposed into singular and regular parts as

$$\psi = 1 + \frac{M_{\rm P}}{2r_{\rm BH}} + \phi\,,\ \Phi = 1 - \frac{M_\Phi}{2r_{\rm BH}} + \eta, \tag{3.26}$$

[4] The conformal flatness in the PN approximation holds in the 1PN order in the harmonic coordinates [25].

where we assume that the puncture is located at $x^i_{\rm P}$ and $r_{\rm BH} = |x^i - x^i_{\rm P}|$ is a coordinate distance from the puncture. Here, $M_{\rm P}$ is a freely specifiable constant called the bare mass, and M_Φ is another freely specifiable constant. We determine the value of $M_{\rm P}$ to obtain a desired value of the (irreducible) mass of the BH. The value of M_Φ is determined by the condition in which the ADM and Komar mass agree, which holds in the stationary, asymptotically flat spacetime [28, 29]. The conformally-weighted extrinsic curvature is also decomposed into singular and regular parts as

$$\hat{A}_{ij} = \hat{A}^{\rm BH}_{ij} + \hat{A}^{\rm NS}_{ij}. \tag{3.27}$$

The former part is given by the so-called Bowen-York extrinsic curvature [30],

$$\begin{aligned}\hat{A}^{\rm BH}_{ij} =& \frac{3}{2r^2_{\rm BH}}\left[\bar{P}^{\rm BH}_i \hat{x}_j + \bar{P}^{\rm BH}_j \hat{x}_i - (f_{ij} - \hat{x}_i \hat{x}_j)\bar{P}^{\rm BH}_k \hat{x}^k\right] \\ &+ \frac{3}{r^3_{\rm BH}}\left[\mathring{\varepsilon}_{kil}\bar{S}^l_{\rm BH}\hat{x}^k \hat{x}_j + \mathring{\varepsilon}_{kjl}\bar{S}^l_{\rm BH}\hat{x}^k \hat{x}_i\right],\end{aligned} \tag{3.28}$$

where $\hat{x}^i \equiv (x^i - x^i_{\rm P})/r_{\rm BH}$. $\bar{P}^{\rm BH}_i$ and $\bar{S}^i_{\rm BH}$ are constants,[5] which correspond to linear and spin angular momenta of the puncture, respectively. The index of $\hat{x}^i$ is raised and lowered by the flat metric, f_{ij}. The value of $\bar{P}^{\rm BH}_i$ is determined so that the linear momentum of the binary vanishes, and the value of $\bar{S}^i_{\rm BH}$ is determined to obtain a desired value of the BH spin angular momentum. The regular part, $\hat{A}^{\rm NS}_{ij}$, is decomposed as in the same way done in the CTT approach using an auxiliary vector field $W^{\rm NS}_i$, and assumed to have no transverse-traceless part as

$$\hat{A}^{\rm NS}_{ij} = \mathring{D}_i W^{\rm NS}_j + \mathring{D}_j W^{\rm NS}_i - \frac{2}{3} f_{ij}\mathring{D}_k W^k_{\rm NS}. \tag{3.29}$$

The equation to determine $W^{\rm NS}_i$ is obtained by the momentum constraint as in the same manner as the CTT approach, using the fact that the Bowen-York extrinsic curvature is a homogeneous solution of the conformally-transformed momentum constraint generated by a vector

$$W^{\rm BH}_i = -\frac{1}{4r}\left[7\bar{P}^{\rm BH}_i + \hat{x}_i \hat{x}^j \bar{P}^{\rm BH}_j\right] + \frac{1}{r^2}\mathring{\varepsilon}_{ijk}\hat{x}^j \bar{S}^k_{\rm BH}. \tag{3.30}$$

Finally, we obtain eight elliptic equations to determine $(\phi, \beta^i, \eta, W^i_{\rm NS})$ as

$$\mathring{D}^2\phi = -\frac{1}{8}\psi^{-7}\hat{A}_{ij}\hat{A}^{ij} - 2\pi\psi^5\rho_{\rm H}, \tag{3.31}$$

---

[5] Here the "constant" means that values of Cartesian components do not depend on the position. Formally, $\mathring{D}_i \bar{P}^{\rm BH}_j = \mathring{D}_i \bar{S}^j_{\rm BH} = 0$.

$$\mathring{D}^2\beta^i + \frac{1}{3}\mathring{D}^i\mathring{D}_j\beta^j = 2\hat{A}^{ij}\mathring{D}_j\left(\Phi\psi^{-7}\right) + 16\pi\Phi\psi^3 j^i, \tag{3.32}$$

$$\mathring{D}^2\eta = \frac{7}{8}\Phi\psi^{-8}\hat{A}_{ij}\hat{A}^{ij} + 2\pi\Phi\psi^4(\rho_{\rm H} + 2S), \tag{3.33}$$

$$\mathring{D}^2 W_i^{\rm NS} + \frac{1}{3}\mathring{D}_i\mathring{D}_j W^j_{\rm NS} = 8\pi\psi^6 j_i. \tag{3.34}$$

These equations are solved with outer boundary conditions derived by the asymptotic flatness,[6] i.e.,

$$\phi|_\infty,\ \beta^i\Big|_\infty,\ \eta|_\infty,\ W_i^{\rm NS}\Big|_\infty = 0. \tag{3.35}$$

In contrast to the case of the excision method [31–33], we do not need to (and cannot) impose inner boundary conditions on the BH horizon.

3.3 Hydrostatic Equilibria

To perform dynamical simulations of the binary merger containing the fluid, the initial condition is desirable to be in a hydrostatic equilibrium. Namely, we want to develop a formalism to compute the hydrostatic configuration to obtain the initial data of the fluid variables $(\rho, \varepsilon, h, u^\mu)$. In this section, we describe the formulation to obtain hydrostatic configurations with an irrotational velocity field adopted in this study [34].

3.3.1 *The Hydrostatics in a Spacetime*

To compute hydrostatic configurations, the state of the fluid flow have to be specified a priori. According to our assumption, basic equations for the hydrostatics of a compact binary in the quasiequilibrium state are derived from the condition of irrotation, or the vanishing of the vorticity two-form [35–38],

$$\begin{aligned}\omega_{\mu\nu} &\equiv \left(g_\mu{}^\alpha + u_\mu u^\alpha\right)\left(g_\nu{}^\beta + u_\nu u^\beta\right)\left(\nabla_\alpha u_\beta - \nabla_\beta u_\alpha\right)\\ &= h^{-1}[\nabla_\mu(hu_\nu) - \nabla_\nu(hu_\mu)]\\ &= 0,\end{aligned} \tag{3.36}$$

where we used the local energy-momentum conservation equation,

$$u^\alpha\nabla_\alpha(hu_\mu) + \nabla_\mu h = 0, \tag{3.37}$$

[6] In fact, the boundary condition on β^i at spatial infinity can be $\beta^i|_\infty = \mathring{\varepsilon}^{ijk}\Omega_j x_k$, if we solve the equations in the corotating frame of a binary. However, this does not influence the result in the conformal flatness approximation, because the shift vector always appears in the equations with a differentiated form, and this rotational part always vanishes.

which is obtained by inserting (1.35), (1.33), and (2.43) into (1.34). In the helically symmetric spacetime with a Killing vector field ξ^i, the specific momentum of the fluid hu_μ should be conserved along the direction of this symmetry, and therefore we obtain

$$\begin{aligned}\pounds_\xi(hu_\mu) &= \xi^\alpha\nabla_\alpha(hu_\mu) - \xi^\alpha\nabla_\mu(hu_\alpha) + \nabla_\mu(hu_\alpha\xi^\alpha) \\ &= 0. \end{aligned} \tag{3.38}$$

Here, the sum of first and second terms of this equation vanishes in the irrotational flow due to (3.36). Hence, we obtain the first integral of relativistic Euler equation,

$$hu_\mu\xi^\mu = -C, \tag{3.39}$$

where C is the constant, the value of which we determine to obtain a desired baryon rest mass of the NS. This equation is regarded as the equation to determine the specific enthalpy, h, and all other thermodynamical quantities are obtained by h using the cold EOS. Next, we find that the condition of irrotation, (3.36), implies the existence of a velocity potential Ψ, which determines the four velocity by

$$u_\mu = h^{-1}\nabla_\mu\Psi. \tag{3.40}$$

The velocity potential is determined by an elliptic-type equation derived from the continuity equation as

$$\rho\nabla^\mu\nabla_\mu\Psi + h\nabla^\mu\left(\frac{\rho}{h}\right)\nabla_\mu\Psi = 0. \tag{3.41}$$

3.3.2 The Hydrostatics in the Initial Value Problem

In order to solve the equations of hydrostatics, (3.39) and (3.41), as a part of the initial value problem in general relativity, we perform the 3+1 splitting of the four velocity [34]. First, recall that the helical Killing vector is written as

$$\xi^\mu = \alpha n^\mu + \beta^\mu_{\rm rot}, \tag{3.42}$$

$$\beta^\mu_{\rm rot} = \beta^\mu + \Omega(\partial_\varphi)^\mu. \tag{3.43}$$

We introduce an corotating observer of the binary, whose four velocity v^μ is parallel to the helical Killing vector and decomposed in a 3+1 manner as

$$v^\mu = \Gamma_0(n^\mu + U_0^\mu), \; n_\mu U_0^\mu = 0. \tag{3.44}$$

Comparing this to the helical Killing vector, it turns out that

$$v^\mu = \frac{\Gamma_0}{\alpha}\xi^\mu\,,\ \Gamma_0 = \frac{\alpha}{\sqrt{\alpha^2 - \beta_i^{\rm rot}\beta_{\rm rot}^i}} = \frac{1}{\sqrt{1-U_i^0 U_0^i}}\,,\ U_0^i = \frac{\beta_{\rm rot}^i}{\alpha}. \tag{3.45}$$

Notice that $\Gamma_0 = -n_\mu v^\mu$ denotes the Lorentz factor of the corotating observer seen from the Eulerian observer. The four velocity of the fluid, u^μ, is decomposed in two different ways as

$$u^\mu = \Gamma(v^\mu + V^\mu)\ (\text{where } v_\alpha V^\alpha = 0) \tag{3.46}$$

$$= \Gamma_n(n^\mu + U^\mu)\ (\text{where } n_\alpha U^\alpha = 0), \tag{3.47}$$

where these two expressions correspond to ones seen from the corotating frame and from the asymptotic inertial frame (or the Eulerian observer).[7] By computing $-v_\mu u^\mu$, it is shown that

$$\Gamma = \Gamma_0\Gamma_n(1 - U_i U_0^i), \tag{3.48}$$

$$V^\mu = \frac{\Gamma_0}{1-U_iU_0^i}\left[(U_i - U_i^0)U_0^i n^\mu + (1 - U_i^0 U_0^i)U^\mu - (1 - U_iU_0^i)U_0^\mu\right], \tag{3.49}$$

and therefore these quantities are derived by the quantities in the asymptotic inertial frame of the binary. We now derive the explicit 3+1 form of the hydrostatic equations. The first integral of the relativistic Euler equation, (3.39), becomes

$$h\alpha\frac{\Gamma}{\Gamma_0} = C, \tag{3.50}$$

and this equation gives the specific enthalpy from geometric variables and the the three velocity of the fluid. The three velocity is obtained by projecting the definition of the velocity potential onto the initial hypersurface, and the Lorentz factor of the fluid seen from the Eulerian observer is obtained by the normalization condition as

$$U_i = \frac{1}{h\Gamma_n}D_i\Psi\,,\ \Gamma_n = \sqrt{1 + \frac{1}{h^2}(D^i\Psi)(D_i\Psi)}. \tag{3.51}$$

The equation to determine the velocity potential is obtained by decomposing (3.41), and it derives

$$\rho D^i D_i\Psi + (D_i\Psi)(D^i\rho)$$
$$= h\Gamma_n U_0^i D_i\rho + \rho\left[(D_i\Psi)\left\{D^i \ln\left(\frac{h}{\alpha}\right)\right\} + hU_0^i D_i\Gamma_n\right] + \rho h\Gamma_n K, \tag{3.52}$$

[7] The Lorentz factor Γ_n is essential the same as w used in other chapter of this thesis.

where we used the relation derived by the helical symmetry, $\xi^\mu \nabla_\mu(\text{scalar}) = 0$. The conformal transformation of this equation leads to the expression for $\hat{\gamma}_{ij} = f_{ij}$ and $K = 0$ as

$$\begin{aligned} &\rho \overset{\circ}{D}{}^2 \psi + \left(\overset{\circ}{D}_i \psi\right)\left(\overset{\circ}{D}{}^i \rho\right) \\ &\quad = \psi^4 h \Gamma_n U_0^i \, \overset{\circ}{D}_i \rho + \rho \left[\left(\overset{\circ}{D}_i \psi\right)\left\{\overset{\circ}{D}{}^i \ln\left(\frac{h}{\alpha\psi^2}\right)\right\} + \psi^4 h U_0^i \, \overset{\circ}{D}_i \Gamma_n\right]. \end{aligned} \tag{3.53}$$

We further rewrite this equation introducing new thermodynamical variable,

$$\zeta \equiv \frac{d\ln(\ln h)}{d\ln\rho}, \quad \overset{\circ}{D}_i \rho = \frac{\rho}{\zeta \ln h} \overset{\circ}{D}_i \ln h, \tag{3.54}$$

and dividing the velocity potential into internal and translational parts as

$$\Psi = \Psi_0 + f_{ij} W^i_{\text{orb}} x^j, \quad W^i_{\text{orb}} \equiv (\psi^4 h \Gamma_n U_0^i)_{\text{center}}, \tag{3.55}$$

from numerical reasons. Finally, we solve

$$\begin{aligned} &\zeta \ln h \, \overset{\circ}{D}{}^2 \Psi_0 + \left[(1 - \zeta \ln h) \, \overset{\circ}{D}{}^i \ln h + \zeta \ln h \, \overset{\circ}{D}{}^i \ln(\alpha\psi^2)\right] \overset{\circ}{D}_i \Psi_0 \\ &\quad = \left[\psi^4 h \Gamma_n U_0^i - W^i_{\text{orb}}\right] \overset{\circ}{D}_i \ln h \\ &\qquad + \zeta \ln h \left[W^i_{\text{orb}} \, \overset{\circ}{D}_i \left\{\ln\left(\frac{h}{\alpha\psi^2}\right)\right\} + \psi^4 h U_0^i \, \overset{\circ}{D}_i \Gamma_n\right]. \end{aligned} \tag{3.56}$$

This elliptic equation to determine Ψ_0, for which the coefficient of the principal part vanishes at the stellar surface, solved with the boundary condition

$$\begin{aligned} &\left[(1 - \zeta \ln h) \, \overset{\circ}{D}{}^i \ln h + \zeta \ln h \, \overset{\circ}{D}{}^i \ln(\alpha\psi^2)\right] \overset{\circ}{D}_i \Psi_0 \\ &\quad = \left[\psi^4 h \Gamma_n U_0^i - W^i_{\text{orb}}\right] \overset{\circ}{D}_i \ln h \\ &\qquad + \zeta \ln h \left[W^i_{\text{orb}} \overset{\circ}{D}_i \left\{\ln\left(\frac{h}{\alpha\psi^2}\right)\right\} + \psi^4 h U_0^i \, \overset{\circ}{D}_i \Gamma_n\right], \end{aligned} \tag{3.57}$$

at the stellar surface.[8]

3.4 Free Parameters

Our formalism for computing a quasiequilibrium binary contains free parameters,

$$M_{\text{P}}, \; M_{\Phi}, \; \bar{P}^i_{\text{BH}}, \; \bar{S}^i_{\text{BH}}, \; C, \; \Omega, \; x^i_{\text{rot}}, \tag{3.58}$$

[8] In the practical computation, we take a different approach [34].

which have to be determined in an appropriate manner. On the one hand, some of these parameters have to be chosen so that the binary is in an appropriate quasi-circular state. On other other hand, some parameters have to be chosen so that the quasiequilibrium state is a member of a particular quasiequilibrium *sequence*, which is specified by the BH mass, BH spin, and the NS mass. We solve the constraint and hydrostatic equations described in Sect. 3.2 and Sect. 3.3 with an iterative method until a sufficiently convergent solution is obtained, and the free parameters are determined in each step of this iterative procedure. Typically, we stop the iteration when sufficient convergence of the enthalpy field is obtained [1].

3.4.1 Parameters Associated with the BH

The bare mass, $M_{\rm P}$, is determined to obtain a desired value of the irreducible mass of the BH, which is obtained by the proper area of the apparent horizon (AH) of the BH,

$$M_{\rm irr} \equiv \sqrt{\frac{A_{\rm AH}}{16\pi}}, \; A_{\rm AH} = \int_{\mathscr{S}_{\rm AH}} \sqrt{q}\,{\rm d}^2 x, \tag{3.59}$$

where we assume that the AH, $\mathscr{S}_{\rm AH}$, is the intersection of the initial hypersurface, Σ_0, and an isolated horizon, $\mathscr{H}$ (see [39, 40] and references therein for reviews). The desired value of $M_{\rm irr}$ itself is taken to be the irreducible mass of the BH which has desired mass and spin parameters at infinite separation. (The mass and spin parameters are defined below.) Because we do not know the position of the AH in advance in the puncture framework, we have to determine the position of the AH numerically. The AH is defined as a two surface on which the expansion of the outgoing null vector $l^\mu = (n^\mu + s^\mu)/\sqrt{2}$ vanishes. The numerical procedure and the code to find an AH is called the AH finder. In the computation of the quasiequilibrium states, we adopt the method described in [41] for the AH finder. The detail is described in the Appendix C.

The other mass parameter, M_Φ, is determined by the condition that the ADM mass,[9]

$$\begin{aligned} M_{\rm ADM} &\equiv \frac{1}{16\pi}\oint_\infty \gamma^{ij}\gamma^{kl}\left(\mathring{D}_j\gamma_{ik} - \mathring{D}_k\gamma_{ij}\right){\rm d}S_l \qquad (3.60) \\ &= -\frac{1}{2\pi}\oint_\infty \mathring{D}_i\psi{\rm d}S^i \\ &= M_{\rm P} - \frac{1}{2\pi}\oint_\infty \mathring{D}_i\phi{\rm d}S^i, \qquad (3.61) \end{aligned}$$

and the Komar mass,[10]

[9] See the Appendix A for the ADM integral.

[10] Rigorously speaking, the Komar mass is defined only in a spacetime with a timelike Killing vector field at spatial infinity.

$$
\begin{aligned}
M_{\mathrm{K}} &\equiv -\frac{1}{8\pi}\oint_{\infty} \nabla^{\mu} t^{\nu} \mathrm{d}S_{\mu\nu} \\
&= \frac{1}{4\pi}\oint_{\infty} \mathring{D}_i \alpha \mathrm{d}S^i \\
&= \frac{M_{\Phi}}{2} + \frac{M_{\mathrm{P}}}{2} + \frac{1}{4\pi}\oint_{\infty} \left(\mathring{D}_i \eta - \mathring{D}_i \phi\right) \mathrm{d}S^i,
\end{aligned} \tag{3.62}
$$

agree, which holds when the spacetime is stationary and asymptotically flat [28, 29]. Here, it is assumed that the shift vector and extrinsic curvature fall off sufficiently rapidly at spatial infinity so that they do not contribute to the surface integral. Equating the former with the latter, M_{Φ} is determined to be

$$
M_{\Phi} = M_{\mathrm{P}} - \frac{1}{2\pi}\oint_{\infty} \left(\mathring{D}_i \phi + \mathring{D}_i \eta\right) \mathrm{d}S^i. \tag{3.63}
$$

The linear momentum parameter of the puncture, $\bar{P}_i^{\mathrm{BH}}$, is determined so that the total linear momentum of the binary vanishes. The ADM linear momentum is computed by

$$
P_i \equiv \frac{1}{8\pi}\oint_{\infty} (K_i{}^j - K\gamma_i{}^j)\mathrm{d}S_j, \tag{3.64}
$$

and therefore $\bar{P}_i^{\mathrm{BH}}$ is determined to be

$$
\bar{P}_i^{\mathrm{BH}} = -\int_{\Sigma} j_i \psi^6 \mathrm{d}^3 x, \tag{3.65}
$$

using the Gauss' theorem with a little algebra. The ADM angular momentum of the binary may also be defined by[11]

$$
J_i \equiv \frac{1}{16\pi} \mathring{\varepsilon}_{ijk} \oint_{\infty} (x^j K^{kl} - x^k K^{jl})\mathrm{d}S_l. \tag{3.66}
$$

The spin angular momentum of the BH, S_{BH}, is evaluated on the AH, $\mathscr{S}_{\mathrm{AH}}$, according to the isolated horizon framework. On the AH, we define an approximate rotational Killing vector using the method developed in [42] with the normalization condition proposed in [43]. The detail of the computation is described in the Appendix C. We focus only on the case in which the BH spin is aligned or antialigned

[11] The definition of the ADM angular momentum is ambiguous due to a supertranslation freedom in general spacetimes. In the conformal flatness approximation, however, this quantity is defined unambiguously [13, 14].

with the angular momentum of the binary in this study, so that the axis of the BH spin is uniquely determined to be the z axis. Therefore, we only consider the approximate Killing vector ϕ^i associated with the rotation in this direction. Using ϕ^i obtained by this method, the spin angular momentum of the BH $S_{\rm BH} = S_{\rm BH}^{(\phi)}$ is computed via the surface integral at the AH,

$$S_{\rm BH}^{(\phi)} = \frac{1}{8\pi} \int_{\mathscr{S}_{\rm AH}} K_{ij} \phi^i \mathrm{d}S^j . \tag{3.67}$$

We adjust $\bar{S}_{\rm BH}^z$ to obtain a desired value of $S_{\rm BH}$. We note that $\bar{S}_{\rm BH}^z$ and $S_{\rm BH}$ do not agree exactly in the BH–NS binary spacetime due to the contribution to the extrinsic curvature from the NS, associated with W_i.

Because we adopt a conformal flatness approximation for the induced metric, the Christodoulou mass of the BH evaluated on the AH [44],

$$M_{\mathscr{H}} = \sqrt{M_{\rm irr}^2 + \frac{S_{\rm BH}^2}{4M_{\rm irr}^2}}, \tag{3.68}$$

and the gravitational mass evaluated at spatial infinity, $M_{\rm BH}$, do not agree even for a single BH system due to the presence of so-called junk waves. This difference leads to an ambiguity in defining the nondimensional spin parameter of the BH. Here, we define the nondimensional spin parameter of the BH with respect to the mass evaluated *at spatial infinity*, i.e.,

$$a \equiv \frac{S_{\rm BH}}{M_{\rm BH}^2}, \tag{3.69}$$

in a single BH spacetime for given values of $M_{\rm irr}$ and $S_{\rm BH}$. The reason for this is that the mass and nondimensional spin parameter of the BH evaluated at the AH quickly (in our simulations, within $\sim$1 ms) relax to $M_{\rm BH}$ and a, defined at spatial infinity, respectively, as the BH absorbs the junk radiation in the vicinity of the BH [43, 45]. We note that these values show the damping oscillation before the relaxation in the same manner as the "scalar-curvature spin" of [43] shows, because our method of evaluating these values in the simulation is basically the same as the method to define the scalar-curvature spin in [43] (see Sect. 5.2.2).

3.4.2 Parameters Associated with the NS

The constant of the first integral of the Euler equation, C, is determined so that the baryon rest mass of the NS,

$$M_{\rm B} \equiv \int_{\Sigma} \rho u^\mu \mathrm{d}V_\mu$$

$$= \int_{\Sigma} \rho \Gamma_n \psi^6 \mathrm{d}^3 x, \tag{3.70}$$

takes the desired value. The value itself is taken to be the baryon rest mass of the NS which has the desired ADM mass, $M_{\rm NS}$, in isolation.

3.4.3 Parameters of the Binary

In addition to the geometric and hydrodynamic variables, we have to obtain the value of the orbital angular velocity, Ω, and the location of the rotational axis or the center of mass of the binary, $x^i_{\rm rot}$, to specify the binary configuration for a given separation, d. In the following, we assume that the rotational axis is always chosen to be the z axis, and describe the method to determine Ω and the position of each object, which we assume to lie on the x axis. Notice that the position of the NS determines the position of the BH uniquely and vice versa, as long as the value of d is given. We still call this position-fixing procedure as determining the location of the rotational axis, or the center of mass of the binary.

The orbital angular velocity of the binary, Ω, is determined by requiring the force balance at the stellar center, which is defined as the location where the specific enthalpy, h, takes the maximum value along the direction connecting two objects. Namely, the stellar center is the point at which the relation,

$$\left.\frac{\partial \ln h}{\partial x}\right|_{\rm center} = 0, \tag{3.71}$$

holds. Using the logarithm of (3.50), this equation leads to[12]

$$\left(\frac{\partial \ln \alpha}{\partial x} + \frac{\partial \Gamma}{\partial x} \right)_{\rm center} = \left.\frac{\partial \Gamma_0}{\partial x}\right|_{\rm center}, \tag{3.72}$$

and inserting (3.45) and (3.43) into Γ_0 derives

$$\begin{aligned} \left(\frac{\partial \ln \alpha}{\partial x} + \frac{\partial \gamma}{\partial x} \right)_{\rm center} = & \frac{1}{2} \left[1 - \frac{\psi^4}{\alpha^2} \hat{\gamma}_{ij} \left(\beta^i + \Omega x^i_{\rm NS} \right) \left(\beta^j + \Omega x^j_{\rm NS} \right) \right]^{-1} \\ & \times \left[\frac{\partial}{\partial x} \left(\frac{\psi^4}{\alpha^2} \right) \hat{\gamma}_{ij} \left(\beta^i + \Omega x^i_{\rm NS} \right) \left(\beta^j + \Omega x^j_{\rm NS} \right) \right. \\ & \left.\left. + \frac{2\psi^4}{\alpha^2} \hat{\gamma}_{ij} \left(\beta^i + \Omega x^i_{\rm NS} \right) \left\{ \frac{\partial \beta^j}{\partial x} + \Omega (\partial_y)^j \right\} \right] \right|_{\rm center}, \end{aligned} \tag{3.73}$$

[12] In fact, this expression is not complete because Γ also depends on Ω. However, this does not cause any problem as far as we know.

where the stellar center is assumed to be located at $x^i_{\rm NS}$. The angular velocity, Ω, is obtained by solving this quadratic equation.

We have no definite condition to determine the location of the center of mass of the binary in the puncture framework, although it is automatically determined by requiring the total linear momentum of a binary vanishes in the excision method [31–33]. We use this ambiguity to reduce an unphysical initial orbital eccentricity, which is unavoidable in the computation of quasicircular binaries.[13] It was found [1, 48] that orbits with a small eccentricity could be obtained using the "3PN-J method," i.e., a phenomenological method to determine the location of the rotational axis in which the total angular momentum of the binary for a given value of Ωm_0 agrees with that calculated from the third-order PN (3PN) approximation. Specifically, the location of the rotational axis is chosen from the condition that the orbital angular momentum of the binary agrees with a sum of 3PN nonspin terms given in [49] and of 2.5PN spin terms given in [50] (see also [51]) for a given value of Ωm_0. When the spin angular momentum of the BH is parallel to the orbital angular momentum of the binary and the spin angular momentum of the NS is negligible, the PN formula of the orbital angular velocity $J(X)$ is given by

$$\begin{aligned}\frac{J(X)}{m_0^2} = \nu X^{-1/2}\Bigg[& 1 + \left(\frac{3}{2}+\frac{\nu}{6}\right)X + \left(\frac{27}{8}-\frac{19}{8}\nu+\frac{\nu^2}{24}\right)X^2 \\ & + \left\{\frac{135}{16}+\left(-\frac{209323}{5040}+\frac{41}{24}\pi^2\right)\nu+\frac{31}{24}\nu^2+\frac{7}{1296}\nu^3\right\}X^3 \\ & + a\left(-\frac{10}{3}Q\nu-\frac{5}{2}\nu\right)X^{3/2} \\ & + a\left\{\left(-7+\frac{217}{72}\nu\right)Q\nu+\left(-\frac{21}{8}+\frac{35}{12}\nu\right)\nu\right\}X^{5/2}\Bigg], \end{aligned} \tag{3.74}$$

where $X = (\Omega m_0)^{2/3}$ is the PN parameter and $\nu = M_{\rm BH}M_{\rm NS}/m_0^2 = Q/(1+Q)^2$ is the symmetric mass ratio. We found that the angular momentum of the binary tends to be larger if the center of mass is chosen to be closer to the BH, for a given separation, d. For completeness, we also show the PN formula of the binding energy of a binary, $E(X)$, as

$$\begin{aligned}\frac{E(X)}{m_0} = -\frac{\nu}{2}X\Bigg[& 1 + \left(-\frac{3}{4}-\frac{\nu}{12}\right)X + \left(-\frac{27}{8}+\frac{19}{8}\nu-\frac{\nu^2}{24}\right)X^2 \\ & + \left\{-\frac{675}{64}+\left(\frac{209323}{4032}-\frac{205}{96}\pi^2\right)\nu-\frac{155}{96}\nu^2-\frac{35}{5184}\nu^3\right\}X^3 \\ & + a\left(\frac{8}{3}Q\nu+2\nu\right)X^{3/2} \\ & + a\left\{\left(8-\frac{31}{9}\right)Q\nu+\left(3-\frac{10}{3}\nu\right)\nu\right\}X^{5/2}\Bigg]. \end{aligned} \tag{3.75}$$

[13] Some effort to reduce eccentricity is done by adding an approaching velocity [33, 46, 47].

It is observed that the thermodynamic relation for a circular orbit [52],

$$\frac{dE}{d\Omega} = \Omega \frac{dJ}{d\Omega}, \tag{3.76}$$

holds as a result of the balance equations of the energy and orbital angular momentum. It should be noted that the numerical computation of the BH-NS binary with the 3PN-J method seems to give a smaller value of $|E(X)|$ than a value computed by the 3PN approximation.

References

1. K. Kyutoku, M. Shibata, K. Taniguchi, Phys. Rev. D **79**, 124018 (2009)
2. K. Kyutoku, H. Okawa, M. Shibata, K. Taniguchi, Phys. Rev. D **84**, 064018 (2011)
3. LORENE website. http://www.lorene.obspm.fr/
4. G.B. Cook, Living Rev. Relativ. **3**, 5 (2000)
5. L. Bildsten, C. Cutler, Astrophys. J. **400**, 175 (1992)
6. C.S. Kochanek, Astrophys. J. **398**, 234 (1992)
7. S.L. Shapiro, S.A. Teukolsky, *Black Holes, White Dwarfs, and Neutron Stars: The Physics of Compact Objects* (Wiley, New York, 1983)
8. P. Marronetti, S.L. Shapiro, Phys. Rev. D **68**, 104024 (2003)
9. T.W. Baumgarte, S.L. Shapiro, Phys. Rev. D **80**, 064009 (2009)
10. T.W. Baumgarte, S.L. Shapiro, Phys. Rev. D **80**, 899001(E) (2009)
11. W. Tichy, Phys. Rev. D **84**, 024041 (2011)
12. R. Arnowitt, S. Deser, C.W. Misner, in *Gravitation*, ed. by L. Witten (Wiley, New York, 1962)
13. J.W. York, in *Gravitational Radiation*, ed. by L. Smarr (Cambridge University Press, Cambridge, 1979)
14. E. Gourgoulhon, 3+1 formalism in general relativity, Lecture Notes in Physics. vol. 846 (Springer-Verlag Berlin, 2012)
15. M. Alcubierre, *Introduction to* 3+1 *Numerical Relativity* (Oxford science publications, Oxford, 2008)
16. J.W. York, Phys. Rev. Lett. **82**, 1350 (1999)
17. H.P. Pfeiffer, J.W. York, Phys. Rev. D **67**, 044022 (2003)
18. J.W. York, J. Math. Phys. **14**, 456 (1973)
19. J.W. York, Ann. Inst. Henri Poincaré **21**, 319 (1974)
20. S. Brandt, B. Brügmann, Phys. Rev. Lett. **78**, 3606 (1997)
21. M. Campanelli, C.O. Lousto, P. Marronetti, Y. Zlochower, Phys. Rev. Lett. **96**, 111101 (2006)
22. J.G. Baker, J. Centrella, D.I. Choi, M. Koppitz, J. van Meter, Phys. Rev. Lett. **96**, 111102 (2006)
23. J.R. Wilson, G.J. Mathews, Phys. Rev. Lett. **75**, 4161 (1995)
24. J.R. Wilson, G.J. Mathews, P. Marronetti, Phys. Rev. D **54**, 1317 (1996)
25. L. Blancet, G. Faye, Phys. Rev. D **63**, 062005 (2001)
26. M. Shibata, K. Uryū, Phys. Rev. D **74**, 121503(R) (2006)
27. M. Shibata, K. Uryū, Class. Quantum Gravity **24**, S125 (2007)
28. R. Beig, Phys. Lett. A **69**, 153 (1978)
29. A. Ashtekar, A. Magnon-Ashtekar, J. Math. Phys. **20**, 793 (1979)
30. J.M. Bowen, J.W. York, Phys. Rev. D **21**, 2047 (1980)
31. K. Taniguchi, T.W. Baumgarte, J.A. Faber, S.L. Shapiro, Phys. Rev. D **75**, 084005 (2007)
32. K. Taniguchi, T.W. Baumgarte, J.A. Faber, S.L. Shapiro, Phys. Rev. D **77**, 044003 (2008)
33. F. Foucart, L.E. Kidder, H.P. Pfeiffer, S.A. Teukolsky, Phys. Rev. D **77**, 124051 (2008)

34. E. Gourgoulhon, P. Grandclément, K. Taniguchi, J.A. Marck, S. Bonazzola, Phys. Rev. D **63**, 064029 (2001)
35. S. Bonazzola, E. Gourgoulhon, J.A. Marck, Phys. Rev. D **56**, 7740 (1997)
36. H. Asada, Phys. Rev. D **57**, 7292 (1998)
37. M. Shibata, Phys. Rev. D **58**, 024012 (1998)
38. S.A. Teukolsky, Astrophys. J. **504**, 442 (1998)
39. A. Ashtekar, B. Krishnan, Living Rev. Relativ. **7**, 10 (2004)
40. E. Gourgoulhon, J.L. Jaramillo, Phys. Rep. **423**, 159 (2006)
41. L.M. Lin, J. Novak, Class. Quantum Gravity **24**, 2665 (2007)
42. G.B. Cook, B.F. Whiting, Phys. Rev. D **76**, 041501 (2007)
43. G. Lovelace, R. Owen, H.P. Pfeiffer, T. Chu, Phys. Rev. D **78**, 084017 (2008)
44. D. Christodoulou, Phys. Rev. D **25**, 1596 (1970)
45. S. Dain, C.O. Lousto, Y. Zlochower, Phys. Rev. D **78**, 024039 (2008)
46. H.P. Pfeiffer, D.A. Brown, L.E. Kidder, L. Lindblom, G. Lovelace, M.A. Scheel, Class. Quantum Gravity **24**, S59 (2007)
47. K. Kiuchi, Y. Sekiguchi, M. Shibata, K. Taniguchi, Phys. Rev. D **80**, 064037 (2009)
48. M. Shibata, K. Kyutoku, T. Yamamoto, K. Taniguchi, Phys. Rev. D **79**, 044030 (2009)
49. L. Blanchet, Phys. Rev. D **65**, 124009 (2002)
50. L. Blanchet, A. Buonanno, G. Faye, Phys. Rev. D **74**, 104034 (2006)
51. G. Faye, L. Blanchet, A. Buonanno, Phys. Rev. D **74**, 104033 (2006)
52. J.L. Friedman, K. Uryū, M. Shibata, Phys. Rev. D **65**, 064035 (2002)

Chapter 4
Methods of Simulations

Numerical simulations of binary mergers are performed using an adaptive-mesh refinement (AMR) code SACRA [1]. In this chapter, we describe the formulation, the gauge conditions, and the numerical scheme adopted in the code. They are essentially the same as those described briefly in [2–4].

4.1 The BSSN-Puncture Formalism

The $3+1$ formalism described in Chap. 3, it is usually called the ADM formalism [5], seems to be an appropriate tool, at a first glance, to perform numerical simulations as the Cauchy problem of general relativity. However, it is known that the ADM formalism cannot be applied to numerical relativity, because the violation of the constraint, which is unavoidable in numerical simulations due to numerical errors and boundary conditions, increases monotonically and finally crashes the simulation. This pathological behavior is due to the mathematical structure of the formalism and resultant behavior of configuration variables at "off-shell". Specifically, numerical errors do not obey hyperbolic equations so that they cannot propagate to escape, and the errors increase exponentially in the ADM formalism. Several effort has been paid to avoid this constraint violation by reformulating evolution equations by many researchers. To date, one of the most successful formalism in numerical relativity is the so-called Baumgarte-Shapiro-Shibata-Nakamura (BSSN) formalism [6, 7]. This formalism suppresses the constraint violation introducing new variables and constraints simultaneously. More importantly, the momentum constraint is added to the evolution equation to eliminate terms which purely correspond to the gauge mode in the linear regime [8].

While the original BSSN formalism had been fairly successful in studying many problems, such as instabilities of rotating stars and the merger of binary NSs, it could not evolve the spacetime containing the singularity, i.e., the BH, for about ten years after its birth. The evolution of binary BH mergers were first performed

K. Kyutoku, *The Black Hole–Neutron Star Binary Merger in Full General Relativity*, Springer Theses, DOI: 10.1007/978-4-431-54201-8_4,

in a generalized harmonic formulation [9, 10], and soon after this first success two groups succeeded in performing binary BH merger simulations based on the BSSN formalism [11, 12]. One of the important modification is that the formalism developed in [11, 12] evolves the inverse of the conformal factor (up to some positive power) instead of the conformal factor itself or its logarithm, which diverges at the singularity. The other important modification is the choice of the gauge condition, which is now called the moving-puncture gauge. They adopted the 1+log slicing, or the "K-driver" condition for the lapse function [13], and the "Γ-driver" condition for the shift vector [14].

Before describing the BSSN-puncture formalism, we comment on the finite differentiation method adopted in the gravitational part of SACRA. We evaluate the spatial derivative by a fourth-order central finite difference, except for the advection terms with the form $\beta^i \partial_i$(variable), which are evaluated by a fourth-order noncentered, upwind finite difference. To evaluate the fourth-order central differentiation at the jth grid point, the data of $(j-2)$–$(j+2)$th points are required. As for the fourth-order upwind differentiation at jth point, the data of $(j-3)$–$(j+1)$th or $(j-1)$–$(j+3)$th points are required, depending on the windward direction. We employ a fourth-order Runge-Kutta method for the time evolution.

4.1.1 BSSN Variables and Evolution Equations

SACRA [1] solves the Einstein evolution equations in the BSSN formalism with the moving-puncture gauge condition in the Cartesian coordinates.[1] In this formalism, the evolution variables are changed from (γ_{ij}, K_{ij}) to

$$W \equiv \gamma^{-1/6}, \tag{4.1}$$

$$\tilde{\gamma}_{ij} \equiv \gamma^{-1/3}\gamma_{ij}, \tag{4.2}$$

$$\tilde{A}_{ij} \equiv \gamma^{-1/3}\left(K_{ij} - \frac{1}{3}K\gamma_{ij}\right), \tag{4.3}$$

$$K \equiv \gamma^{ij}K_{ij}, \tag{4.4}$$

$$\tilde{\Gamma}^i \equiv \tilde{\gamma}^{jk}\tilde{\Gamma}^i{}_{jk} = -\partial_j\tilde{\gamma}^{ij}, \tag{4.5}$$

where $\tilde{\Gamma}^k_{ij}$ is the Christoffel symbol of $\tilde{\gamma}_{ij}$, and the choice of the conformal factor, W, is taken from [16]. The most important change here is the introduction of the conformal connection functions $\tilde{\Gamma}^i$ as an independent variable. The reason to do this is that the principal part of the conformal Ricci tensor, $\tilde{R}_{ij}$, which is a complicated second-order derivative operator on $\tilde{\gamma}_{ij}$, becomes the Laplacian on $\tilde{\gamma}_{ij}$, and the system

---

[1] The BSSN formalism can be applied to other coordinate systems, such as spherical coordinates [15]. Basically, it is only required to replace ∂_i by $\mathring{D}_i$, and subtract the Christoffel symbol of f_{ij}, which vanishes in the Cartesian coordinates, from the Christoffel symbol of $\tilde{\gamma}_{ij}$, $\tilde{\Gamma}^k{}_{ij}$.

of equations becomes (weakly) hyperbolic [17]. The evolution equations of them are obtained by performing conformal transformation to the ADM evolution equations of (γ_{ij}, K_{ij}), (3.12) and (3.13), and are given by

$$\left(\partial_t - \beta^i \partial_i\right) W = \frac{1}{3} W \left(\alpha K - \partial_i \beta^i\right), \tag{4.6}$$

$$\left(\partial_t - \beta^k \partial_k\right) \tilde{\gamma}_{ij} = -2\alpha \tilde{A}_{ij} + \tilde{\gamma}_{ik} \partial_j \beta^k + \tilde{\gamma}_{jk} \partial_i \beta^k - \frac{2}{3} \tilde{\gamma}_{ij} \partial_k \beta^k, \tag{4.7}$$

$$\begin{aligned}
\left(\partial_t - \beta^k \partial_k\right) \tilde{A}_{ij} = & -W^2 \left(D_i D_j \alpha - \frac{1}{3} \gamma_{ij} D^2 \alpha\right) + W^2 \alpha \left(R_{ij} - \frac{1}{3} R \gamma_{ij}\right) \\
& + \alpha \left(K \tilde{A}_{ij} - 2 \tilde{A}_{ik} \tilde{A}_j{}^k\right) - 8\pi W^2 \alpha \left(S_{ij} - \frac{1}{3} S \gamma_{ij}\right) \\
& + \tilde{A}_{ik} \partial_j \beta^k + \tilde{A}_{jk} \partial_i \beta^k - \frac{2}{3} \tilde{A}_{ij} \partial_k \beta^k,
\end{aligned} \tag{4.8}$$

$$\left(\partial_t - \beta^i \partial_i\right) K = -D^2 \alpha + \alpha \left[\tilde{A}_{ij} \tilde{A}^{ij} + \frac{1}{3} K^2 + 4\pi (\rho_{\mathrm{H}} + S)\right], \tag{4.9}$$

$$\begin{aligned}
\left(\partial_t - \beta^j \partial_j\right) \tilde{\Gamma}^i = & -2 \tilde{A}^{ij} \partial_j \alpha + 2\alpha \left(\tilde{\Gamma}^i_{jk} \tilde{A}^{jk} - \frac{3}{W} \tilde{A}^{ij} \partial_j W - \frac{2}{3} \tilde{\gamma}^{ij} \partial_j K - 8\pi \tilde{\gamma}^{ij} j_j\right) \\
& + \tilde{\gamma}^{jk} \partial_j \partial_k \beta^i + \frac{1}{3} \tilde{\gamma}^{ij} \partial_j \partial_k \beta^k \\
& + \left(\partial_j \tilde{\gamma}^{jk}\right) \partial_k \beta^i - \frac{2}{3} \left(\partial_j \tilde{\gamma}^{ij}\right) \partial_k \beta^k.
\end{aligned} \tag{4.10}$$

Here, the Hamiltonian and momentum constraints are used to derive evolution equations of K and $\tilde{\Gamma}^i$, respectively. Especially, eliminating $\partial_j \tilde{A}^{ij}$, which can always be set to zero by choosing the transverse-traceless gauge in the linear regime, from the evolution equation of $\tilde{\Gamma}^i$ is the essential part of the BSSN formalism. Finally, the Ricci tensor is computed by

$$R_{ij} = \tilde{R}_{ij} + R^W_{ij}, \tag{4.11}$$

$$\begin{aligned}
\tilde{R}_{ij} = & -\frac{1}{2} \tilde{\gamma}^{kl} \partial_k \partial_l \tilde{\gamma}_{ij} + \frac{1}{2} \left(\tilde{\gamma}_{ik} \partial_j \tilde{\Gamma}^k + \tilde{\gamma}_{jk} \partial_i \tilde{\Gamma}^k\right) + \frac{1}{2} \tilde{\Gamma}^{kl}_l \partial_k \tilde{\gamma}_{ij} \\
& + \tilde{\Gamma}^{kl}_i \tilde{\Gamma}_{jkl} + \tilde{\Gamma}^{kl}_j \tilde{\Gamma}_{ikl} + \tilde{\Gamma}^{kl}_i \tilde{\Gamma}_{klj},
\end{aligned} \tag{4.12}$$

$$R^W_{ij} = \frac{1}{W} \tilde{D}_i \tilde{D}_j W + \frac{1}{W} \tilde{\gamma}_{ij} \tilde{D}^2 W - \frac{2}{W^2} \tilde{\gamma}_{ij} \left(\tilde{D}^k W\right) \left(\tilde{D}_k W\right), \tag{4.13}$$

where $\tilde{\Gamma}^{ij}_k \equiv \tilde{\gamma}^{jl} \tilde{\Gamma}^i_{kl}$. In this decomposition, $\tilde{R}_{ij}$ and R^W_{ij} approximately represent gravitational-wave and gravitational-potential parts, respectively. As is described above, the principal part of $\tilde{R}_{ij}$ is the Laplacian of $\tilde{\gamma}_{ij}$ thanks to the introduction of $\tilde{\Gamma}^i$. Throughout these computations, $\tilde{\Gamma}^i$ appears on the right-hand side only when it is differentiated, and never used as it is. Namely, the first derivative of the conformal metric, such as $\partial_j \tilde{\gamma}^{ij}$ and $\tilde{\Gamma}^{kl}_l$, is always computed from $\tilde{\gamma}_{ij}$. This trick makes the numerical simulation stable.

4.1.2 BSSN Constraints

In the BSSN formalism, the Hamiltonian and momentum constraints, (3.10) and (3.11), are also subjected to the conformal transformation. In addition, new constraint equations arise as the price of increasing new independent variables. The BSSN constraint equations are given by

$$\mathscr{H} \equiv R + \frac{2}{3}K^2 - \tilde{A}_{ij}\tilde{A}^{ij} - 16\pi\rho_{\mathrm{H}} = 0, \tag{4.14}$$

$$\mathscr{M}_i \equiv \tilde{\gamma}^{jk}\tilde{D}_k\tilde{A}_{ij} - \frac{3}{W}\tilde{A}_i^j\tilde{D}_jW - \frac{2}{3}\tilde{D}_iK - 8\pi j_i = 0, \tag{4.15}$$

$$\tilde{\gamma} = f = 1, \tag{4.16}$$

$$\tilde{\gamma}^{ij}\tilde{A}_{ij} = 0, \tag{4.17}$$

$$\tilde{\Gamma}^i = -\partial_j\tilde{\gamma}^{ij}. \tag{4.18}$$

Among these constraints, we explicitly force the algebraic constraints (4.16) and (4.17) at all the time step, including the intermediate steps of the Runge-Kutta time evolution, by recomputing the variables as

$$\tilde{\gamma}_{ij} = \tilde{\gamma}^{-1/3}\tilde{\gamma}_{ij}, \tag{4.19}$$

$$W = \tilde{\gamma}^{-1/6}W, \tag{4.20}$$

$$\tilde{A}_{ij} = \tilde{\gamma}^{-1/3}\tilde{A}_{ij} - \frac{1}{3}\left(\tilde{\gamma}^{kl}\tilde{A}_{kl}\right)\left(\tilde{\gamma}^{-1/3}\tilde{\gamma}_{ij}\right), \tag{4.21}$$

$$K = K + \tilde{\gamma}^{ij}\tilde{A}_{ij}, \tag{4.22}$$

where the left-hand side is the new values of the variables computed by the right-hand side expression using the old values of the variables. The values of (γ_{ij}, K_{ij}) do not change before and after this recomputation. Notice that these equations are all satisfied automatically if the BSSN algebraic constraints are satisfied. The Hamiltonian and momentum constraints, $\mathscr{H}$ and $\mathscr{M}_i$, are monitored during the simulation to measure the degree of the constraint violation.

The BSSN formalism is called the free-evolution scheme, because the Einstein constraint equations are not solved in the evolution. This is justified, because it can be shown that the constraints are satisfied during the evolution as long as the initial condition satisfy the constraints and the evolution is governed by the Einstein evolution equations thanks to the Bianchi identity. The ADM formalism also shares this feature, but it cannot halt the unstable growth of slight but unavoidable constraint violation, in contrast to the BSSN formalism. There exists other formulation in which some (typically, only the Hamiltonian constraint) or all of the Einstein constraint equations are solved simultaneously, and they are called the partially-constrained and fully-constrained scheme [15, 18].

4.1.3 The Moving Puncture Gauge Condition

Introducing an auxiliary variable B^i and a parameter η_s, which we typically set to be[2] $\sim 1/M_{BH}$ in units of $G = c = M_\odot = 1$, SACRA employs the moving-puncture gauge in the form [20]

$$\left(\partial_t - \beta^j \partial_j\right) \alpha = -2\alpha K, \tag{4.23}$$

$$\left(\partial_t - \beta^j \partial_j\right) \beta^i = \frac{3}{4} B^i, \tag{4.24}$$

$$\left(\partial_t - \beta^j \partial_j\right) B^i = \left(\partial_t - \beta^j \partial_j\right) \tilde{\Gamma}^i - \eta_s B^i. \tag{4.25}$$

We always give initial data of the gauge variables by

$$\alpha = W, \beta^i = 0, B^i = 0, \tag{4.26}$$

and never use the value given by the XTCS approach. In particular, replacing α is important, because the lapse function computed in the puncture framework always becomes negative around the puncture (approximately, inside the AH), and makes the the simulation unstable. The motion of the puncture is obtained by integrating the shift vector as

$$\frac{\mathrm{d}x^i_{\mathrm{P}}}{\mathrm{d}t} = -\beta^i, \tag{4.27}$$

because the conformal factor, W, vanishes at the puncture [21], and therefore (4.6) becomes

$$\frac{\partial W}{\partial t} - \beta^i \frac{\partial W}{\partial x^i} = 0, \tag{4.28}$$

which means that the characteristic velocity of the puncture is $-\beta^i$ [11].

4.2 Hydrodynamic Evolution Equations

In computational fluid dynamics, it is fairly advantageous to formulate evolution equations in a conservative form [22, 23], i.e.,

$$\partial_t \mathbf{Q} + \partial_i \mathbf{F}^i(\mathbf{Q}) = \mathbf{S}(\mathbf{Q}), \tag{4.29}$$

---

[2] This parameter η_s has a dimension of the inverse of the time. Because this parameter introduces a time scale to the gauge evolution equation, we cannot choose the time step size arbitrarily even if the CFL condition is satisfied [19].

for a state vector[3] **Q**. If the numerical simulations are performed in the conservative form throughout, conservation of "conserved quantities", such as the total rest mass, is ensured up to the truncation error. Although this does not hold if we adopt the AMR technique, in which the interpolation is necessary, the violation of conservation still gives us an estimate of the error associated with the AMR. The conservative form is also advantageous to accurately capture the position of shock waves, which is the most troublesome part of computational fluid dynamics.

The spatial differentiation involved in the source terms, **S**(**Q**), are evaluated by a fourth-order central differentiation as is done for the gravitational part. The time evolution is performed by a fourth-order Runge-Kutta time evolution. Our scheme of evaluating the advection terms, $\partial_i \mathbf{F}^i(\mathbf{Q})$, is described below.

4.2.1 Evolution Equations in a Conservative Form

The hydrodynamic evolution equation is obtained by the continuity equation and the local energy-momentum conservation equation, (1.33) and (1.34), with decomposing (1.34) into the space and time components by projecting onto the hypersurface, $\gamma_i^{\mu}\nabla_\nu T^\nu_\mu$, and onto the four velocity of the Eulerian observer, $n^\mu \nabla_\nu T^\nu_\mu$. These equations give

$$\partial_t \rho_* + \partial_i \left(\rho_* v^i\right) = 0, \tag{4.30}$$

$$\begin{aligned}\partial_t \left(\rho_* \hat{u}_i\right) + \partial_j \left(\rho_* \hat{u}_i v^j + P\alpha\sqrt{\gamma}\delta_i^j\right) &= -\rho_* \left(hw\partial_i\alpha - \hat{u}_j \partial_i \beta^j + \frac{\alpha}{2hw}\hat{u}_j\hat{u}_k\partial_i\gamma^{jk}\right) \\ &\quad + P\partial_i(\alpha\sqrt{\gamma}),\end{aligned} \tag{4.31}$$

$$\begin{aligned}\partial_t \left(\rho_* \hat{e}\right) + \partial_i \left[\rho_* \hat{e} v^i + P\sqrt{\gamma}\left(v^i + \beta^i\right)\right] &= P\alpha\sqrt{\gamma}K - \rho_* \hat{u}_i \gamma^{ij}\partial_j \alpha \\ &\quad + \frac{\rho_*\alpha}{hw}\hat{u}_i\hat{u}_j K^{ij},\end{aligned} \tag{4.32}$$

where we define the Lorentz factor, three velocity, and conserved variables by

$$w \equiv -n_\mu u^\mu = \alpha u^t, \tag{4.33}$$

$$v^i \equiv \frac{\mathrm{d}x^i}{\mathrm{d}t} = \frac{u^i}{u^t}, \tag{4.34}$$

$$\rho_* \equiv \rho\alpha\sqrt{\gamma}u^t, \tag{4.35}$$

$$\hat{u}_i \equiv hu_i, \tag{4.36}$$

$$\hat{e} \equiv h\alpha u^t - \frac{P}{\rho\alpha u^t}. \tag{4.37}$$

[3] Here, the vector means merely a set of variables.

In deriving these equations, it is essential to use (u^t, u_i) rather than (u_t, u^i) except for the definition of v^i, because we have no direct access to the latter quantities when we use the former. Notice that all of the right-hand sides of (4.30), (4.31), and (4.32) become zeros and the equations become those of special relativity, if the spacetime is flat. This means that the rest mass, linear momentum, and energy are the conserved quantities in a flat spacetime, and indeed the rest mass is the conserved quantity even in the curved spacetime. The energy and linear momentum are not considered to be conserved in general relativity, due to the well known absence of the well-defined local gravitational energy and linear momentum.

The advection terms are handled with a high-resolution[4] central scheme proposed by Kurganov and Tadmor [24] with a third-order piecewise parabolic interpolation for the cell reconstruction. To explain this scheme, which is a monotone upwind-central scheme for conservation laws (MUSCL), let us consider the way to solve a one-dimensional, source-free evolution equation in a conservative form,

$$\partial_t \mathbf{Q} + \partial_x \mathbf{F}^x = 0, \tag{4.38}$$

on a uniform grid with the grid size Δx and with the time step size Δt. Because the differential equation does not make sense at discontinuities such as a shock front, numerical schemes should be able to obtain a weak solution which satisfy the conservation law in an integrated form,[5]

$$\int\int (\partial_t \mathbf{Q} + \partial_x \mathbf{F}^x) \mathrm{d}t \mathrm{d}x = 0. \tag{4.39}$$

The solution at the time step $n+1$ and at the grid point j is obtained by taking the integration cell to be $t^n \leq t \leq t^{n+1}$ and $x_{j-1/2} \leq x \leq x_{j+1/2}$, and is written by

$$\mathbf{Q}_j^{n+1} = \mathbf{Q}_j^n - \frac{\Delta t}{\Delta x} \left(\mathbf{F}_{j+1/2}^n - \mathbf{F}_{j-1/2}^n \right), \tag{4.40}$$

where the superscript and subscript indicate the time step and grid point, respectively. Hereafter, we discuss quantities at n to evaluate the right-hand side and omit superscripts.

The important question is how to evaluate the flux at the cell interface, $\mathbf{F}_{j-1/2}$ and $\mathbf{F}_{j+1/2}$. It is well known that using more information from the windward side is essential for the numerical stability, so the upwind scheme is necessary. In the scheme of Kurganov and Tadmor, the flux at $\mathbf{F}_{j-1/2}$ is evaluated in a flux difference splitting (FDS) manner by

---

[4] "High resolution" means that a higher-order accuracy is obtained for the smooth flow, while the discontinuity is captured accurately by lowering the accuracy to a first order around discontinuities. Godunov's theorem states that the scheme must be a first order to capture the discontinuities without invoking instabilities.

[5] In this sense, the conservative scheme of computational fluid dynamics is sometimes considered to be based on a finite volume method.

$$\mathbf{F}_{j-1/2} = \frac{1}{2}\left[\mathbf{F}^{\mathrm{R}}_{j-1/2} + \mathbf{F}^{\mathrm{L}}_{j-1/2} - |a|_{j-1/2}\left(\mathbf{Q}^{\mathrm{R}}_{j-1/2} - \mathbf{Q}^{\mathrm{L}}_{j-1/2}\right)\right], \tag{4.41}$$

where L and R stand for the "left" and "right" state of the cell interface, respectively, and $|a|$ denotes the local propagation velocity. The local propagation velocity, $|a|$, is computed as a maximum of the characteristic velocity which is defined by the eigenvalue of the Jacobian matrix,

$$M_{ab} \equiv \frac{\partial F_a}{\partial Q_b}, \tag{4.42}$$

where (a, b) denotes the components of the flux and state vectors. The eigenvalues are computed both at left and right states, and the maximum value is searched over both states. In general relativistic hydrodynamics, where the five-dimensional state and flux vectors are given by

$$\mathbf{Q} = \begin{pmatrix} \rho_* \\ \rho_* \hat{u}_x \\ \rho_* \hat{u}_y \\ \rho_* \hat{u}_z \\ \rho_* \hat{e} \end{pmatrix}, \mathbf{F}^i = \begin{pmatrix} \rho_* v^i \\ \rho_* \hat{u}_x v^i + P\alpha\sqrt{\gamma}\delta^i_x \\ \rho_* \hat{u}_y v^i + P\alpha\sqrt{\gamma}\delta^i_y \\ \rho_* \hat{u}_z v^i + P\alpha\sqrt{\gamma}\delta^i_z \\ \rho_* \hat{e} v^i + P\sqrt{\gamma}(v^i + \beta^i) \end{pmatrix}, \tag{4.43}$$

eigenvalues of the advection in i-direction are given by [25] triple roots v^i and

$$\lambda^i_\pm = \frac{v^i\alpha^2(1-c_{\mathrm{s}}^2) - \beta^i c_{\mathrm{s}}^2(\alpha^2 - V^2) \pm \alpha c_{\mathrm{s}}\sqrt{(\alpha^2 - V^2)[\gamma^{ii}(\alpha^2 - V^2 c_{\mathrm{s}}^2) - (1-c_{\mathrm{s}}^2)V^i V^i]}}{|\alpha^2 - V^2 c_{\mathrm{s}}^2|}, \tag{4.44}$$

where $V^i \equiv v^i + \beta^i$, $V^2 \equiv \gamma_{ij} V^i V^j$, and c_{s} is the sound velocity defined by

$$c_{\mathrm{s}}^2 \equiv \frac{1}{h}\left[\left(\frac{\partial P}{\partial \rho}\right)_\varepsilon + \frac{P}{\rho^2}\left(\frac{\partial P}{\partial \varepsilon}\right)_\rho\right]. \tag{4.45}$$

It is noted that the derivatives for the sound velocity are computed as

$$\left(\frac{\partial P}{\partial \rho}\right)_\varepsilon = K_i \Gamma_i \frac{\Gamma_i - \Gamma_{\mathrm{th}}}{\Gamma_i - 1}\rho^{\Gamma_i - 1} + (\varepsilon - \varepsilon_{i-1})(\Gamma_{\mathrm{th}} - 1), \tag{4.46}$$

$$\frac{1}{\rho}\left(\frac{\partial P}{\partial \varepsilon}\right)_\rho = \Gamma_{\mathrm{th}} - 1, \tag{4.47}$$

with our EOS described in Sect. 2.4. Hence, the solution is obtained when the values of physical quantities at left an right states are specified, using the local propagation velocity determined as

$$|a| \equiv \max\left[\left|\lambda^{\mathrm{L}}_\pm\right|, \left|\lambda^{\mathrm{R}}_\pm\right|, \left|v^{\mathrm{L}}\right|, \left|v^{\mathrm{R}}\right|\right], \tag{4.48}$$

where we come back to the one-dimensional problem and omit superscript, i.

To determine the right state at $j-1/2$ and the left state at $j+1/2$ to high accuracy, we adopt a third-order piecewise-parabolic interpolation around j for

$$\rho_*, \hat{u}_i, \hat{e}, \varepsilon, \frac{P}{\rho}, \left(\frac{\partial P}{\partial \rho}\right)_\varepsilon, \frac{1}{\rho}\left(\frac{\partial P}{\partial \varepsilon}\right)_\rho, \tag{4.49}$$

whereas the geometric variables at $j \pm 1/2$ are computed by a straightforward arithmetic mean of the values at j and $j \pm 1$. In the piecewise-parabolic interpolation, the physical distribution of quantities are assumed to take a Taylor-expanded form,

$$Q(x) = Q(x_j) + (x - x_j)\frac{\partial Q(x_j)}{\partial x} + \frac{1}{2}(x - x_j)^2\frac{\partial^2 Q(x_j)}{\partial x^2}, \tag{4.50}$$

and the numerical value at x_j is considered to be the cell-average of this distribution function. Therefore, the distribution function is rewritten using the value at j as

$$Q(x) = \left[Q_j - \frac{1}{24}\left(\frac{\partial^2 Q}{\partial x^2}\right)_j (\Delta x)^2\right] + (x - x_j)\left(\frac{\partial Q}{\partial x}\right)_j + \frac{1}{2}(x - x_j)^2\left(\frac{\partial^2 Q}{\partial x^2}\right)_j. \tag{4.51}$$

Evaluating Q_j, $(\partial Q/\partial x)_j$, and $(\partial Q^2/\partial x^2)_j$ by the second-order, central finite differentiation, we finally obtain

$$Q^{\mathrm{L}}_{j+1/2} = Q_j + \frac{1}{3}\Delta Q_{j+1/2} + \frac{1}{6}\Delta Q_{j-1/2}, \tag{4.52}$$

$$Q^{\mathrm{R}}_{j-1/2} = Q_j - \frac{1}{6}\Delta Q_{j+1/2} - \frac{1}{3}\Delta Q_{j-1/2}, \tag{4.53}$$

$$\Delta Q_{j+1/2} \equiv Q_{j+1} - Q_j, \tag{4.54}$$

$$\Delta Q_{j-1/2} \equiv Q_j - Q_{j-1}. \tag{4.55}$$

In order to avoid (possibly unphysical) local extrema at the cell interface, we further modify this expression using a minmod limiter defined by

$$\mathrm{minmod}(a, b) = \begin{cases} a, & (b > a > 0, 0 > a > b) \\ b, & (a > b > 0, 0 > b > a) \\ 0, & (a > 0 > b, b > 0 > a) \end{cases} \tag{4.56}$$

as

$$Q^{\mathrm{L}}_{j+1/2} = Q_j + \frac{1}{3}\mathrm{minmod}\left(\Delta Q_{j+1/2}, b\Delta Q_{j-1/2}\right) + \frac{1}{6}\mathrm{minmod}\left(\Delta Q_{j-1/2}, b\Delta Q_{j+1/2}\right), \tag{4.57}$$

$$Q^{\mathrm{R}}_{j-1/2} = Q_j - \frac{1}{6}\mathrm{minmod}\left(\Delta Q_{j+1/2}, b\Delta Q_{j-1/2}\right)$$

$$-\frac{1}{3}\mathrm{minmod}\left(\Delta Q_{j-1/2}, b\Delta Q_{j+1/2}\right), \tag{4.58}$$

so that the total variation diminishing (TVD) condition is satisfied. In SACRA, the parameter b is taken to be 3. In a uniform grid, this parameter has to satisfy $1 < b < 4$ to avoid local extrema, and low and high values of b tend to lead dissipative and unstable schemes, respectively.

4.2.2 Recovery of Primitive Variables

In the course of numerical simulations, we have to determine primitive variables (ρ, u_i, ε) consistently from the conserved variables $(\rho_*, \hat{u}_i, \hat{e})$ and geometric variables at each time step. This procedure requires the determination of the Lorentz factor, w, which is accomplished with the aid of the EOS and the normalization condition of the four velocity, $u_\mu u^\mu = -1$. Considering the enthalpy, h, as a function of the Lorentz factor, w, the normalization condition indicates

$$f(w) \equiv 1 + \gamma^{ij}\hat{u}_i\hat{u}_j h(w)^{-2} - w^2 = 0. \tag{4.59}$$

What we have to do here is to find a physically appropriate root, w, of this equation. This is done by an iterative, Newton-Raphson method with an initial guess of w, which we take to be a value at the previous step. Because the rest-mass density, specific enthalpy, and specific internal energy are given by

$$\rho = \frac{\rho_*}{w\sqrt{\gamma}}, \tag{4.60}$$

$$h = \frac{\hat{e}}{w} + \frac{P\sqrt{\gamma}}{\rho_* w}, \tag{4.61}$$

$$\varepsilon = \frac{\hat{e}}{w} + \frac{P\sqrt{\gamma}}{\rho_* w} - 1 - \frac{Pw\sqrt{\gamma}}{\rho_*}, \tag{4.62}$$

the pressure is also determined as a function of w. Specifically, we obtain using our EOS described in Sect. 2.4,

$$h(w)^{-1} = [\Gamma_{\rm th}(w^2-1)+1] \times \left[\Gamma_{\rm th} w\hat{e} + K_i \frac{\Gamma_i - \Gamma_{\rm th}}{\Gamma_i - 1}\left(\frac{\rho_*}{w\sqrt{\gamma}}\right)^{\Gamma_i - 1} - (\Gamma_{\rm th} - 1) - \varepsilon_{i-1}(\Gamma_{\rm th} - 1)\right]^{-1}, \tag{4.63}$$

and now we solve (4.59) using this relation to obtain the correct value of w, and hence (ρ, u_i, ε) and h until sufficient convergence is obtained.

4.2.3 An Artificial Atmosphere

Because the vacuum is not allowed in any conservative hydrodynamic scheme, we put an artificial atmosphere of a small density outside the NS. Specifically, we choose a typical rest-mass density of the atmosphere $\rho_{\min}$ to be $10^{-9}\rho_{\max}$, where $\rho_{\max}$ is the maximum rest-mass density in the NS at an initial instant. In typical situations, $\rho_{\max} \approx 10^{15}$ gcm^{-3} and therefore $\rho_{\min} \approx 10^{6}$ gcm^{-3}. In the actual computation, we further introduce a typical radius of the atmosphere $R_{\rm crit} \approx 150$–200 km, and put an atmosphere determined by

$$\rho_* = \exp\left[\min\left(1 - \frac{r}{R_{\rm crit}}, 0\right)\right]\rho_{\min}, \tag{4.64}$$

if ρ_* becomes zero or negative during the evolution. The conserved velocity, $\hat{u}_i$, is set to be zero, and the conserved energy density, $\hat{e}$, is set to be

$$\hat{e} = 1 + \varepsilon_{\rm cold}(\rho_{\min}). \tag{4.65}$$

The total rest mass of the atmosphere is always less than $10^{-4}\,M_\odot$ (in many situations, less than $10^{-5}\,M_\odot$), and hence, we can safely neglect spurious effects by accretion of the atmosphere onto the remnant disk as far as the disk mass is much larger than $10^{-4}\,M_\odot$.

4.3 Adaptive Mesh Refinement

The computational difficulty in numerical relativity is that the problem usually involves two different length scales. In dynamical simulations of the compact binary coalescence, on the one hand, we have to resolve both compact objects with a sufficient grid resolution to perform accurate simulations. The radius of the NS is typically ≈10–15 km, and the least required resolution is ≈250–400 m to obtain convergent result by experience. On the other hand, we have to extend the computational domain to the local wave zone of the initial quasiequilibrium state, because the usual boundary condition and extracting technique of gravitational waves are both become accurate only at local wave zones. A typical gravitational wavelength at an initial instant is written using the initial angular velocity Ω_0 as

$$\begin{aligned}\lambda_0 &\equiv \frac{\pi c}{\Omega_0} = \frac{Gm_0}{c^2}\frac{\pi}{G\Omega_0 m_0/c^3}\\ &\approx 250(1+Q)\left(\frac{M_{\rm NS}}{1.35M_\odot}\right)\left(\frac{0.025}{G\Omega_0 m_0/c^3}\right)\ \text{km},\end{aligned} \tag{4.66}$$

where we inserted G and c for clarity. If we are willing to cover the whole computational domain by the same resolution adapted to the compact object, the total grid number becomes as large as $\approx 10^{11}$. Because $\approx$100–150 variables with double precision (8 Bytes) are required for the Runge-Kutta time evolution of general relativistic hydrodynamic simulations, the total required memory will be $\approx$100 Tbyte and unachievable.

An important fact is that such a high resolution is not required except for the region in the vicinity of compact objects. Actually, $\gtrsim$10 grid points are adequate to resolve gravitational waves, whose wavelength is no less than $\approx$100 km even just before the merger. Therefore, it is becoming a standard technique to use an adaptive mesh refinement (AMR) technique in numerical relativity.[6]

4.3.1 The Grid Structure

SACRA adopts the AMR technique [1] of Berger-Oliger type [26] (see also [20]). Figure 4.1 depicts the schematic view of the AMR algorithm implemented in SACRA, which we describe below. The AMR grids consist of a number of computational domains, each of which has uniform, nvertex-centered Cartesian grids with $(2N + 1, 2N + 1, N + 1)$ for (x, y, z) with the equatorial symmetry at $z = 0$ imposed. The AMR grids are classified into two categories. One is a coarser domain, which covers a wide region, including both the BH and NS, with its origin fixed at the approximate center of mass throughout the simulation. The other is a finer domain, two sets of which comove with compact objects and cover the region in the vicinity of these objects. We denote the edge length of the largest domain, the number of the coarser domains, and the number of the finer domains by $2L$, $l_{\rm c}$, and $2l_{\rm f}$, respectively. Namely, the total number of the domains is $l_{\rm c}+2l_{\rm f}$. The grid spacing for each domain is $h_l = L/(2^l N)$, where $l = 0$–$(l_{\rm c}+l_{\rm f}-1)$ is the depth of each domain. The size of a time step $(\Delta t)_l$ is determined by h_l in each domain with the Courant-Friedrich-Levy (CFL) factor, which we set to be 0.5 in this study. Hence, sizes of the time step of two neighboring domains differ by a factor of 2, and a sub-cycling in time is necessary. Namely, the time evolution of the lth domain is performed twice during single time evolution of the $(l-1)$th domain. Figure 4.1 shows this sub-cycling structure in time. During the time evolution of $(l-1)$th domain once with the time step Δt, the lth domain is evolved twice in time with $\Delta t/2$. Similarly, the $(l+1)$th domain is evolved four times in time with $\Delta t/4$. The ordering of this sub-cycling is determined recursively so that each domain is maximally evolved as long as it does not precede the domain located just above it.[7] The number (1)–(7) in Fig. 4.1 shows the time ordering for the specific configuration shown in the figure. Exceptionally, $(\Delta t)_{l<l_{\rm c}}$ in coarser domains are chosen to be the same as $(\Delta t)_{l=l_{\rm c}}$ of the coarsest finer domains in our simulations.

[6] The other solution to the difficulty with different length scales is to use non-uniform grids.

[7] In other words, the domain never precede any of the domains located above it.

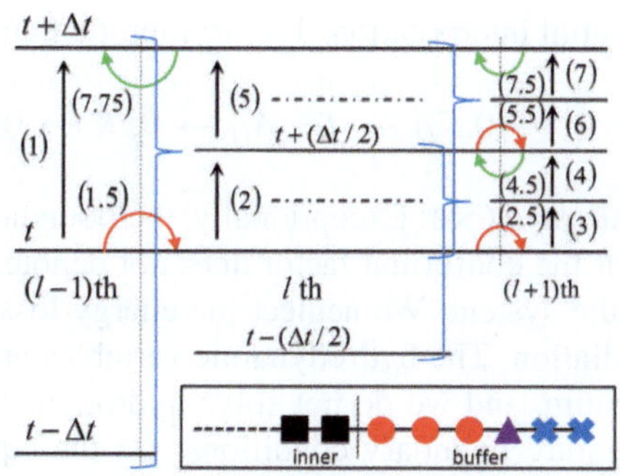

Fig. 4.1 The schematic figure of the AMR algorithm implemented in SACRA. This example shows the evolution from t to $t + \Delta t$ for $(l - 1)$th (*left*), lth (*middle*), and $(l + 1)$th (*right*) domains, where Δt is the time step determined by the CFL number in the $(l - 1)$th domain. The coordinate time proceeds upward, and the *solid horizontal line* denotes each computational grid at each time. The *dot-dashed horizontal line* denotes the intermediate step of the Runge-Kutta time evolution. Numbers in brackets denotes the ordering of the procedure. The inset at the *right bottom* shows the configuration of the buffer zone, which is denoted by colored points, i.e., the *red circles*, *purple triangle*, and *blue crosses*. The inner zone is denoted by *black squares*. The data of all buffer points are given by the fifth-order Lagrange interpolation using the data of the domain located just above it, when the time agrees with that of the domain located just above it (Step (1.5), (2.5), and (5.5) in this figure, shown by *red, curved arrows*). The *red circles* are evolved using the same method as for the inner zone during two Runge-Kutta evolution. This means that the spatial interpolation is not performed at $t + (\Delta t/2)$, at which $(l - 1)$th domain does not have the data, for the *red circles* of the lth domain. The *purple triangle* is also evolved in a similar way with the exception that it is allowed to evaluate advection terms by a second-order, upwind finite difference when the windward side is the outside. The data of *blue crosses* are given by the Lagrange interpolation of fifth- and second-order in space and time, respectively, at all the time steps including the intermediate steps of the Runge-Kutta evolution. In this example, this interpolation is performed at $t + (\Delta t/2)$ and all the intermediate steps of (2) and (5) for the lth domain using the data at $t - \Delta t$, t, and $t + \Delta t$ of the $(l - 1)$th domain. After the catching-up in time with the domain located just above it, the better-resolved data of the domain are populated to the domain located just above it (Step (4.5), (7.5), and (7.75) in this figure, shown by *green, curved arrows*)

4.3.2 Boundary Conditions and Data in the Buffer Zone

For $l = 0$ domain, we put two boundary points in both positive and negative directions. Values of metric quantities at these boundary points are determined by the outgoing-wave boundary condition at all the time steps including the intermediate steps of the Runge-Kutta time evolution. The outgoing boundary condition is written by

$$F(t, r) = [F(t - \Delta t, r - \Delta r) - F(\infty)]\frac{r - \Delta r}{r} + F(\infty), \tag{4.67}$$

where r denotes the distance from the coordinate origin and $\Delta r = \Delta t$ in the unit of $c = 1$, for a variable F which obeys the wave equation. The value of $F(t-\Delta t, r-\Delta r)$

is determined by an spatial interpolation. The asymptotic flatness condition,

$$\alpha \to 1,\, \beta^i \to 0,\, B^i \to 0,\, \tilde{\gamma}_{ij} \to f_{ij},\, \tilde{A}_{ij} \to 0,\, K \to 0,\, \tilde{\Gamma}^i \to 0, \tag{4.68}$$

is imposed for the value of $F(\infty)$. Exceptionally, the boundary value of W is fixed because the $1/r$ part of the conformal factor does not denote the wave component, but the total mass of the system. We neglect the energy loss from the system due to the gravitational radiation. The hydrodynamic variables at $l = 0$ domain are set to be those of the vacuum, and we do not solve hydrodynamics at $l = 0$ domain. Hence, they do not require boundary conditions. On the equatorial plane, we do not impose the outgoing-wave boundary condition, but the equatorially-symmetric boundary condition. Specifically,

$$\begin{aligned} &F(-z) = F(z),\, F^A(-z) = F^A(z),\, F^z(-z) = -F^z(z) \\ &F_{AB}(-z) = F_{AB}(z),\, F_{Az}(-z) = -F_{Az}(z),\, F_{zz}(-z) = F_{zz}(z), \end{aligned} \tag{4.69}$$

where (A, B) denote (x, y) components. This equatorially symmetric condition is also applied to all the $l \neq 0$ computational domains.

All the other computational domains have six buffer points in both positive and negative directions. The schematic configuration of the buffer zone is shown in the inset at the right bottom of Fig. 4.1. The issues described in this paragraph are also explained in the caption of Fig. 4.1 for the specific grid configuration shown in the figure. In these domains, values at the buffer zone at the beginning of each time step are computed using the fifth-order Lagrange interpolation in space from values of the domain located just above them, when the coordinate time agrees with that of the domain located just above them. In other words, the values are not interpolated at the beginning of the "second" Runge-Kutta time evolution. Exceptionally, we switch to lower-order interpolation to avoid numerical instability when the value of $(\rho_*, \hat{e}, h)$ becomes lower than that of the atmosphere. During the Runge-Kutta evolution, inner three of six buffer points are evolved using the same method as in the inner zone, i.e., true computational grids. The fourth point is also evolved, but the advection terms are allowed to be evaluated by a second-order, upwind finite differentiation when the windward side is the outside. It should be recalled that the fourth-order upwind scheme requires three windward points, but the fourth point has only two outer points. At the outer two of six buffer points, the variables are not evolved, and instead the values of the variables are computed using the Lagrange interpolation, which is fifth- and second-order in space and time, respectively, at the beginning of the "second" Runge-Kutta time evolution and all the intermediate steps of the Runge-Kutta time evolution. The interpolation in time is changed to a first-order one for hydrodynamic variables to avoid numerical instability if the value oscillates in time. After the two Runge-Kutta evolution, the data computed in the domain with a better resolution are populated to the domain located just above it, which has a lower resolution.

References

1. T. Yamamoto, M. Shibata, K. Taniguchi, Phys. Rev. D **78**, 064054 (2008)
2. K. Kyutoku, M. Shibata, K. Taniguchi, Phys. Rev. D **82**, 044049 (2010)
3. K. Kyutoku, M. Shibata, K. Taniguchi, Phys. Rev. D **84**, 049902(E) (2011)
4. K. Kyutoku, H. Okawa, M. Shibata, K. Taniguchi, Phys. Rev. D **84**, 064018 (2011)
5. R. Arnowitt, S. Deser, C.W. Misner, in *Gravitation*, ed by L. Witten (Wiley, New York, 1962), pp. 227–265
6. M. Shibata, T. Nakamura, Phys. Rev. D **52**, 5428 (1995)
7. T.W. Baumgarte, S.L. Shapiro, Phys. Rev. D **59**, 024007 (1998)
8. T. Nakamura, K. Oohara, Y. Kojima, Prog. Theor. Phys. Suppl. **90**, 1 (1987)
9. F. Pretorius, Phys. Rev. Lett. **95**, 121101 (2005)
10. F. Pretorius, Class. Quantum Gravity **23**, S529 (2006)
11. M. Campanelli, C.O. Lousto, P. Marronetti, Y. Zlochower, Phys. Rev. Lett. **96**, 111101 (2006)
12. J.G. Baker, J. Centrella, D.I. Choi, M. Koppitz, J. van Meter, Phys. Rev. Lett. **96**, 111102 (2006)
13. C. Bona, J. Massó, E. Seidel, J. Stela, Phys. Rev. Lett. **75**, 600 (1995)
14. M. Alcubierre, B. Brügmann, P. Diener, M. Koppitz, D. Pollney, E. Seidel, R. Takahashi, Phys. Rev. D **67**, 084023 (2003)
15. E. Gourgoulhon, 3+1 formalism in general relativity, Lecture Notes in Physics. vol. 846 (Springer-Verlag Berlin, 2012)
16. P. Marronetti, W. Tichy, B. Brügmann, J.A. González, U. Sperhake, Phys. Rev. D **77**, 064010 (2008)
17. M. Alcubierre, *Introduction to 3+1 Numerical Relativity* (Oxford science publications, Oxford, 2008)
18. S. Bonazzola, E. Gourgoulhon, P. Grandclément, J. Novak, Phys. Rev. D **70**, 104007 (2004)
19. E. Schnetter, Class. Quantum Gravity **27**, 167001 (2010)
20. B. Brügmann, J.A. González, M. Hannam, S. Husa, U. Sperhake, W. Tichy, Phys. Rev. D **77**, 024027 (2008)
21. S. Brandt, B. Brügmann, Phys. Rev. Lett. **78**, 3606 (1997)
22. J.M. Marti, E. Müller, Living Rev. Relativity **6**, 7 (2003)
23. J.A. Font, Living Rev. Relativity **11**, 7 (2008)
24. A. Kurganov, E. Tadmor, J. Comp. Phys. **160**, 241 (2000)
25. F. Banyuls, J.A. Font, J.M. Ibáñez, J.M. Marti, J.A. Miralles, Astrophys. J. **476**, 221 (1997)
26. M.J. Berger, J. Oliger, J. Comp. Phys. **53**, 484 (1984)

Chapter 5
Diagnostics for Numerical Simulations

The purpose of this study is to analyze gravitational waves and merger remnants, and clarify their dependence on the NS EOS. In this chapter, we describe methods to extract gravitational waves from numerically computed spacetimes within a finite domain, and to obtain values of physical quantities associated with the remnant disk and BH.

5.1 Gravitational Waves

Identification of gravitational-wave contents in a numerically-computed spacetime is not a trivial task. For example, the metric $g_{\mu\nu}$ constructed from $(\alpha, \beta^i, \gamma_{ij})$ does not directly related to the gauge-invariant, transverse-traceless gauge quantities even in the weak field. Several extraction technique has been proposed, and comparison among different extraction techniques have been performed by several authors [1–3]. The simplest method to extract gravitational waves is to use the quadrupole formula, (1.3). In particular, this is the only method to compute gravitational waves in Newtonian simulations. However, the definition of the mass quadrupole moment is ambiguous in general relativity. One of successful methods is to adopt the gauge-invariant, Regge-Wheeler-Zerilli-Moncrief formalism [4–6] at a finite coordinate radius, but this method assumes a specific background in the far region, and also sometimes introduce unphysical high-frequency noise [3]. In this study, gravitational waves are extracted using the Newman-Penrose quantity [7–9], or the Weyl scalar Ψ_4, at a finite coordinate radius.

We note that the most reliable method so far may be the Cauchy-Characteristic-Extraction (CCE) technique [3, 10], in which quantities associated with gravitational waves are evolved from the finite extraction radius to future null infinity as a post process. This method is more preferable than other methods in that it has no systematic errors associated with the finite extraction radius, but it cannot be implemented easily (and we do not), because the CCE technique requires a different evolution code based on the Bondi-Sachs coordinates [11, 12].

K. Kyutoku, *The Black Hole–Neutron Star Binary Merger in Full General Relativity*, Springer Theses, DOI: 10.1007/978-4-431-54201-8_5,

5.1.1 Extracting Gravitational Waves

The Weyl tensor, which is defined in the 4-dimensional spacetime as

$$C_{\alpha\mu\beta\nu} \equiv {}^4R_{\alpha\mu\beta\nu} - \frac{1}{2}\left(g_{\alpha\beta}{}^4R_{\mu\nu} - g_{\alpha\nu}{}^4R_{\beta\mu} - g_{\beta\mu}{}^4R_{\alpha\nu} + g_{\mu\nu}{}^4R_{\alpha\beta}\right)$$
$$- \frac{1}{6}{}^4R(g_{\alpha\beta}g_{\mu\nu} - g_{\alpha\nu}g_{\beta\mu}), \quad (5.1)$$

contains information of the spacetime curvature irrelevant to the Ricci tensor,[1] especially gravitational waves. The Weyl tensor is decomposed into "electric" and "magnetic" parts [13], which are defined by

$$E_{\mu\nu} \equiv C_{\alpha\mu\beta\nu}n^\alpha n^\beta, \quad (5.2)$$
$$B_{\mu\nu} \equiv C^*_{\alpha\mu\beta\nu}n^\alpha n^\beta, \quad (5.3)$$
$$C^*_{\alpha\mu\beta\nu} \equiv \frac{1}{2}C_{\alpha\mu\lambda\sigma}\varepsilon^{\lambda\sigma}{}_{\beta\nu} \quad (5.4)$$

for an observer n^μ, as

$$C_{\alpha\mu\beta\nu} = p_{\alpha\beta}E_{\mu\nu} - p_{\alpha\nu}E_{\beta\mu} - p_{\beta\mu}E_{\alpha\nu} + p_{\mu\nu}E_{\alpha\beta}$$
$$- n_\alpha B_{\mu\lambda}\varepsilon^\lambda{}_{\beta\nu} + n_\mu B_{\alpha\lambda}\varepsilon^\lambda{}_{\beta\nu} - n_\beta B_{\nu\lambda}\varepsilon^\lambda{}_{\alpha\mu} + n_\nu B_{\beta\lambda}\varepsilon^\lambda{}_{\alpha\mu}, \quad (5.5)$$

where $p_{\mu\nu} = g_{\mu\nu} + 2n_\mu n_\nu$. Using symmetry properties of the Weyl tensor, which are the same as those of the Riemann tensor, it is shown that the electric and magnetic parts are symmetric, trace-free, and spatial, i.e.,

$$E_{\mu\alpha}n^\alpha = B_{\mu\alpha}n^\alpha = 0, \quad (5.6)$$

which implies that these two tensors are computed on a spatial hypersurface of constant time. Specifically, we obtain

$$E_{ij} = R_{ij} + KK_{ij} - K_i{}^kK_{jk} - \frac{16\pi}{3}\rho_{\rm H}\gamma_{ij} - 4\pi\left[S_{ij} - \frac{1}{3}S\gamma_{ij}\right], \quad (5.7)$$
$$B_{ij} = \varepsilon_i{}^{mn}[D_mK_{nj} - 4\pi\gamma_{jm}j_n], \quad (5.8)$$

using the Einstein equations.

Gravitational waves are extracted by calculating the outgoing part of the Weyl scalar Ψ_4 [7–9], which is defined from the Weyl tensor using null tetrads

[1] In general relativity, the Ricci tensor has the same information as that of the energy-momentum tensor.

$(l^\mu, k^\mu, m^\mu, \bar{m}^\mu)$ by[2]

$$\Psi_4 \equiv -C_{\alpha\mu\beta\nu}\bar{m}^\alpha k^\mu \bar{m}^\beta k^\nu \tag{5.9}$$

at finite coordinate radii $r \approx 100m_0$ assuming that the spacetime is the vacuum. We normalize the outgoing and ingoing null vectors as

$$l^\mu = \frac{1}{\sqrt{2}}(n^\mu + s^\mu), \tag{5.10}$$

$$k^\mu = \frac{1}{\sqrt{2}}(n^\mu - s^\mu), \tag{5.11}$$

and the complex null tetrad m^μ is chosen to be

$$m^\mu = \frac{1}{\sqrt{2}}\left[(e_{\hat{\theta}})^\mu + \mathrm{i}\,(e_{\hat{\varphi}})^\mu\right], \tag{5.12}$$

where $(e_{\hat{\theta}})^\mu$ and $(e_{\hat{\varphi}})^\mu$ are the normalized coordinate bases of θ and φ, respectively.[3] After decomposing the Weyl tensor into the electric and magnetic parts, Ψ_4 is found to be evaluated by

$$\Psi_4 = -(E_{ij} - \mathrm{i}B_{ij})\bar{m}^i\bar{m}^j. \tag{5.13}$$

We always decompose Ψ_4 into modes associated with a spin-weighted spherical harmonics of spin weight -2 as

$$\Psi_4(r,\theta,\varphi) = \sum_{l,m} \Psi_4^{lm}(r)\,{}_{-2}Y_{lm}(\theta,\varphi), \tag{5.14}$$

where ${}_sY_{lm}(\theta,\varphi)$ is defined using the Wigner's d-function d^l_{ms} by

$${}_sY_{lm}(\theta,\varphi) = (-1)^s\sqrt{\frac{2l+1}{4\pi}}d^l_{m(-s)}(\theta)\mathrm{e}^{\mathrm{i}m\varphi}, \tag{5.15}$$

---

[2] The signature of the Weyl scalar is not universal. Some authors use the opposite signature, and then it is compensated by an additional minus sign for the relation between Ψ_4 and gravitational waves.

[3] The Gram-Schmidt orthogonalization is not performed in SACRA. It is justified because the orthogonality is satisfied at the limit of the infinite radius, where the weak field limit is achieved.

$$d^l_{ms}(\theta) = \sum_{k=\max(0,m-s)}^{\min(l+m,l-s)} (-1)^k \frac{\sqrt{(l+m)!(l-m)!(l+s)!(l-s)!}}{(l+m-k)!(l-s-k)!k!(k+s-m)!}$$
$$\times \cos^{2l+m-s-2k}\left(\frac{\theta}{2}\right) \sin^{2k+s-m}\left(\frac{\theta}{2}\right), \tag{5.16}$$

by projecting on a sphere with a constant coordinate radius. In this study, we extract all the $l \le 4$ and $-l \le m \le l$ modes. Notice that $\pm m$ modes contain the same information in the equatorially symmetric spacetime.

5.1.2 The Fixed-Frequency Integration Method

Because the Weyl tensor is equivalent to the Riemann tensor in the vacuum, the gravitational waveform is obtained by integrating Ψ_4 twice in time as

$$h_+(t) - \mathrm{i}h_\times(t) = \int^t \int^{t'} \Psi_4(t'')\mathrm{d}t''\mathrm{d}t', \tag{5.17}$$

with appropriate choices of two integration constants. Here, gravitational-wave components are defined using the deviation of the metric from the background, flat metric $h_{\mu\nu} \equiv g_{\mu\nu} - \eta_{\mu\nu}$ by

$$h_+ \equiv \frac{1}{2}(h_{\hat{\theta}\hat{\theta}} - h_{\hat{\varphi}\hat{\varphi}}), \tag{5.18}$$
$$h_\times \equiv h_{\hat{\theta}\hat{\varphi}} = h_{\hat{\varphi}\hat{\theta}}, \tag{5.19}$$

in the orthonormal basis, and the relation in the traceless-transverse gauge and weak-field limit,

$$R_{titj} = -R_{tirj} = R_{rirj} = -\frac{1}{2}\frac{\partial^2 h_{ij}}{\partial t^2}, \tag{5.20}$$

is assumed. However, it is difficult to know appropriate values of the integration constants. To make the matter worse, it is well known that a straightforward evaluation of (5.17) usually introduces unphysical drift components, which may be ascribed to the random noise in numerical simulations [14]. In our previous works [15],[4] we perform the straightforward integration of $\Psi_4(t)$, and then subtract a quadratic function of the form $a_2t^2 + a_1t + a_0$ to fix the integration constants and eliminate

[4] We also performed this direct time integration in [16, 17], and present refined results computed by the fixed-frequency integration method, with which the results changes only very slightly for nonspinning BH–NS binaries.

unphysical drift components in the waveform, using the least-square fitting to obtain coefficients a_0, a_1, and a_2.

In this study, we adopt a fixed-frequency integration method proposed in [14] to obtain gravitational waveforms with less unphysical components. In this method, we first perform a Fourier transformation of Ψ_4 as

$$\tilde{\Psi}_4(\omega) = \int \Psi_4(t) \mathrm{e}^{-\mathrm{i}\omega t} \mathrm{d}t. \tag{5.21}$$

Using this, (5.17) is rewritten as

$$h_+(t) - \mathrm{i}h_\times(t) = -\frac{1}{2\pi} \int \frac{\tilde{\Psi}_4(\omega)}{\omega^2} \mathrm{e}^{\mathrm{i}\omega t} \mathrm{d}\omega. \tag{5.22}$$

We then replace $1/\omega^2$ of the integrand with $1/\omega_0^2$ for $|\omega| < \omega_0$, where ω_0 is a positive free parameter in this method. Namely, the gravitational waveform is computed as

$$h_+(t) - \mathrm{i}h_\times(t) = -\frac{1}{2\pi} \int \frac{\tilde{\Psi}_4(\omega)}{\max[\omega^2, \omega_0^2]} \mathrm{e}^{\mathrm{i}\omega t} \mathrm{d}\omega. \tag{5.23}$$

By appropriately choosing ω_0, this procedure works as a high-pass filter, and suppresses only unphysical, low-frequency drift components of gravitational waves. In the simulation of the binary merger, the lowest frequency of physical gravitational waves is naturally determined by the initial orbital angular frequency of the binary, Ω_0. As proposed in [14], we choose ω_0 to be $\sim 0.8\, m\Omega_0$ for $m \neq 0$ mode gravitational waves, where m is the azimuthal quantum number. For the $m = 0$ mode gravitational waves, we adopt $\omega_0 \sim 0.8\, \Omega_0$ and confirm that our results depend only very weakly on this choice.

We also adopt this method to calculate the energy ΔE and angular momentum ΔJ in the z direction radiated by gravitational waves. These quantities are computed by integrating the emission rate, which is written in terms of the Weyl scalar as [18, 19]

$$\frac{\mathrm{d}E}{\mathrm{d}t} = \lim_{r\to\infty} \frac{r^2}{16\pi} \oint_{\mathscr{S}} \left| \int \Psi_4 \mathrm{d}t \right|^2 \mathrm{d}\Omega, \tag{5.24}$$

$$\frac{\mathrm{d}J}{\mathrm{d}t} = -\lim_{r\to\infty} \frac{r^2}{16\pi} \oint_{\mathscr{S}} \mathrm{Re}\left[\left(\int \bar{\Psi}_4 \mathrm{d}t \right) \left(\int\int \partial_\varphi \Psi_4 \mathrm{d}t \mathrm{d}t' \right) \mathrm{d}\Omega \right], \tag{5.25}$$

where $\mathscr{S}$ is a coordinate sphere of the radius r and $\mathrm{d}\Omega = \sin\theta \mathrm{d}\theta \mathrm{d}\varphi$. Explicit expressions for these quantities in terms of mode coefficients are given in the Appendix B. Although these rates must be evaluated at infinity, we alternatively use values computed at a finite coordinate radius. The error associated with the finite radius effect is estimated to be less than $\approx$5 %, and the error associated with the finite grid resolution seems to be larger.[5] We can also compute the kick velocity of the remnant BH by integrating the emission rate of the linear momentum,

$$\frac{dP_i}{dt} = \lim_{r\to\infty} \frac{r^2}{16\pi} \oint_{\mathscr{S}} s_i \left| \int \Psi_4 \mathrm{d}t \right|^2 \mathrm{d}\Omega, \tag{5.26}$$

but we do not investigate the kick velocity in detail, because the kick velocity necessarily involves the mode couplings between different (l, m) modes, and numerical errors are large. This is also the case for x and y components of the orbital angular velocity.

5.1.3 The Taylor-T4 Formula

To prepare the gravitational-wave templates for the actual data analysis, it is most important to accurately model the evolution of the gravitational-wave frequency [20]. In principle, the orbital evolution of a circular binary in the inspiral phase, at which the PN, adiabatic approximation is valid, is computed by

$$\frac{\mathrm{d}\Omega}{\mathrm{d}t} = \frac{\mathscr{L}}{\mathrm{d}E/\mathrm{d}\Omega}, \tag{5.27}$$

where the orbital binding energy $E(X)$ and the gravitational-wave luminosity $\mathscr{L}(X)$ are given by polynomials of a (square root of a) nondimensional angular velocity $X(t) = [\Omega(t)m_0]^{2/3}$ as the PN parameter, with some logarithmic terms of X. Therefore, numerical relativity waveforms should agree with those predicted by the PN approximation solving this equation in the early inspiral phase. However, it is known that the orbital evolution depends strongly on how this equation is actually solved [21, 22]. To perform meaningful comparisons, it is necessary to choose carefully the practical method to compute the orbital evolution.

For comparisons between numerically calculated gravitational waveforms and those calculated in the PN approximation, we adopt the Taylor-T4 formula for two-point masses in circular orbits [21, 23, 24], which shows remarkable coincidence with numerical relativity results for the nonspinning BH binaries, with an additional contribution from the BH spin angular momentum [25]. In this formula, the time evolution of the orbital angular velocity $\Omega(t)$ and orbital phase $\Phi(t)$ are computed using X by expanding $\mathscr{L}/(\mathrm{d}E/\mathrm{d}X)$ into a single polynomial,

[5] To reduce systematic errors associated with this issue, extrapolation of $\Psi_4(r)$ should be performed assuming some functional form, e.g., $\Psi_4(r) = \Psi_4(r \to \infty) + \sum_{n>1} \Psi_{4,n} r^{-n}$.

$$
\begin{aligned}
\frac{\mathrm{d}X}{\mathrm{d}t} =& \frac{64\nu X^5}{5m_0}\left[1 - \left(\frac{743}{336} + \frac{11}{4}\nu\right)X + \left(4\pi - \frac{113}{12}\chi + \frac{19}{6}\nu a\right)X^{3/2}\right.\\
&+ \left(\frac{34103}{18144} + 5\chi^2 + \frac{13661}{2016}\nu + \frac{59}{18}\nu^2\right)X^2\\
&- \left\{\left(\frac{4159}{672} + \frac{189}{8}\nu\right)\pi + \left(\frac{31571}{1008} - \frac{1165}{24}\nu\right)\chi + \frac{3}{4}\chi^3\right.\\
&\left. - \left(\frac{21863}{1008}\nu - \frac{79}{6}\nu^2\right)a\right\}X^{5/2}\\
&+ \left\{\frac{16447322263}{139708800} - \frac{1712}{105}\gamma_E + \frac{16}{3}\pi^2 - \left(\frac{56198689}{217728} - \frac{451}{48}\pi^2\right)\nu\right.\\
&+ \frac{541}{896}\nu^2 - \frac{5605}{2592}\nu^3 - \frac{856}{105}\ln(16X) - \frac{80\pi}{3}\chi\\
&\left. + \left(\frac{64153}{1008} - \frac{457}{36}\nu\right)\chi^2 + \left(\frac{20}{3}\pi - \frac{1135}{36}\chi\right)\nu a\right\}X^3\\
&- \left\{\left(\frac{4415}{4032} - \frac{358675}{6048}\nu - \frac{91495}{1512}\nu^2\right)\pi\right.\\
&+ \left(\frac{2529407}{27216} - \frac{845827}{6048}\nu + \frac{41551}{864}\nu^2\right)\chi - 12\pi\chi^2 + \left(\frac{1505}{24} + \frac{\nu}{8}\right)\chi^3\\
&\left.\left. - \left(\frac{1580239}{54432} - \frac{451597}{6048}\nu^2 + \frac{2045}{432}\nu^3 + \frac{107}{6}\nu\chi^2\right)a\right\}X^{7/2}\right],
\end{aligned} \tag{5.28}
$$

and the orbital phase are computed by integrating the orbital frequency,

$$
\frac{\mathrm{d}\Phi}{\mathrm{d}t} = \frac{X^{3/2}}{m_0}, \tag{5.29}
$$

where $\nu = Q/(1+Q)^2$, $\chi \equiv aQ/(1+Q)$, and $\gamma_E \approx 0.5772$ is the Euler constant. After $X(t)$ and $\Phi(t)$ are obtained, we calculate the complex gravitational-wave amplitude h^{22} of the $(l, m) = (2, 2)$ mode and the spectrum[6] up to the 3PN order using the formula shown in [25, 26]. Here, h^{22} is

$$
\begin{aligned}
h^{22} =& -8\sqrt{\frac{\pi}{5}}\frac{\nu m_0}{D}\mathrm{e}^{-2\mathrm{i}\Phi}X\left[1 - \left(\frac{107}{42} - \frac{55}{42}\nu\right)X + \left(2\pi - \frac{4}{3}\chi + \frac{2}{3}\nu a\right)X^{3/2}\right.\\
&- \left(\frac{2173}{1512} + \frac{1069}{216}\nu - \frac{2047}{1512}\nu^2\right)X^2 - \left\{\left(\frac{107}{21} - \frac{34}{21}\nu\right)\pi + 24\mathrm{i}\nu\right\}X^{5/2}
\end{aligned}
$$

[6] It is noted that the spectrum at high frequency for the Taylor-T4 formula depends on the location where time-domain gravitational waves are truncated. It does not affect the fitting procedure described in Chaps. 6 and 7 as far as we truncate the waveform before X and the amplitude become too large.

$$+\left\{\frac{27027409}{646800}-\frac{856}{105}\gamma_E+\frac{2}{3}\pi^2-\frac{428}{105}\ln(16X)\right.$$
$$\left.-\left(\frac{278185}{33264}-\frac{41}{96}\pi^2\right)\nu-\frac{20261}{2772}\nu^2+\frac{114635}{99792}\nu^3+\frac{428}{105}\mathrm{i}\pi\right\}X^3\Bigg], \quad (5.30)$$

where D is the distance between the center of mass of the binary and an observer.

Another way for deriving an approximate waveform is to employ an effective one-body (EOB) approach. In the EOB approach, the PN dynamics of two-point masses are mapped onto the geodesic motion of a point particle in a Schwarzschild-like spacetime by comparing the Hamiltonian (see [27] and references therein for reviews). By a suitable resummation technique (the Padé approximation) with calibrating free parameters using numerical-relativity results for the binary BH merger, the dynamics of a binary is tracked in a nonadiabatic manner beyond the ISCO, where the traditional PN approximation completely breaks down. The inspiral-plunge waveform obtained by the EOB approach is matched to the merger-ringdown waveform obtained by the BH perturbation technique (see [28] and references therein for reviews) using values of the mass and spin of the remnant BH obtained by numerical-relativity simulations. In [29], comparisons between numerical waveforms computed in this study and those of the EOB approach are extensively performed.

5.2 Quantities of Merger Remnants

In this study, the material outside the AH is identified with the remnant disk after the merger, and the quantities computed on the AH approximately represent the properties of the BH, which should be defined through the event horizon in principle. The location of the AH is always computed numerically during the simulation by solving $\Theta_{(l)} = 0$ adopting the method described in the Appendix of [30].

5.2.1 Quantities of the Remnant Disk

To estimate the mass of the remnant disk, we calculate the total rest mass outside the AH

$$M_{r>r_{\rm AH}} \equiv \int_{r>r_{\rm AH}} \rho_* \mathrm{d}^3x, \quad (5.31)$$

where $r_{\rm AH} = r_{\rm AH}(\theta, \varphi)$ is the radius of the AH as a function of the angular coordinates. We possibly underestimated disk masses because some of the material escapes from our computational domains and we cannot follow their return which would occur if they are bounded. We also evaluate the total angular momentum of the material located outside the AH, $J_{r>r_{\rm AH}}$, which is approximately defined by

$$J_{r>r_{\rm AH}} \equiv \int_{r>r_{\rm AH}} \rho_* \hat{u}_\varphi {\rm d}^3x, \tag{5.32}$$

where

$$\hat{u}_\varphi = x\hat{u}_y - y\hat{u}_x. \tag{5.33}$$

5.2.2 Quantities of the Remnant Black Hole

We determine key quantities of the remnant BH, i.e., the mass $M_{\rm BH,f}$ and nondimensional spin parameter $a_{\rm f}$, from the circumferential radius of the AH, assuming that the deviation from the Kerr spacetime is negligible in the vicinity of a BH horizon. We estimate the remnant BH mass, $M_{\rm BH,f}$, from the circumferential radius of the AH along the equatorial plane $C_{\rm e}$ divided by 4π, i.e., $C_{\rm e}/4\pi$, which gives the BH mass in the stationary vacuum BH spacetime. Similarly, the nondimensional spin parameter of the remnant BH, $a_{\rm f}$, is estimated from the ratio of the circumferential radius of the AH along the meridional plane $C_{\rm p}$ to $C_{\rm e}$ using the relation [31]

$$\frac{C_{\rm p}}{C_{\rm e}} = \frac{\sqrt{2\hat{r}_+}}{\pi} E\left(\frac{a_{\rm f}^2}{2\hat{r}_+}\right). \tag{5.34}$$

This also holds for the stationary vacuum BH with the nondimensional spin parameter $a_{\rm f}$, and estimating the BH spin by $C_{\rm p}/C_{\rm e}$ is sometimes called the great circle method. Here, $\hat{r}_+ = 1+\sqrt{1-a_{\rm f}^2}$ is a normalized radius of the horizon, and $E(z)$ is an elliptic integral

$$E(z) = \int_0^{\pi/2} \sqrt{1 - z\sin^2\theta}{\rm d}\theta. \tag{5.35}$$

For comparison, the nondimensional spin parameter of the remnant BH is also estimated from $C_{\rm e}$ and the irreducible mass of the remnant BH, $M_{\rm irr,f}$, using the relation

$$M_{\rm irr,f} = \frac{C_{\rm e}}{4\sqrt{2}\pi}\sqrt{1+\sqrt{1-a_{\rm f}{}^2}}, \tag{5.36}$$

which holds for the stationary vacuum BH. The spin parameter obtained using this relation is referred to as $a_{\rm f2}$ according to [15]. Finally, we also estimate $a_{\rm f}$ from the values of the remnant BH computed using approximate conservation laws

$$M_{\mathrm{BH,c}} \equiv M_0 - M_{r>r_{\mathrm{AH}}} - \Delta E, \tag{5.37}$$

$$J_{\mathrm{BH,c}} \equiv J_0 - J_{r>r_{\mathrm{AH}}} - \Delta J. \tag{5.38}$$

Here, we assume that the orbital angular momentum of the BH is negligible. The nondimensional spin parameter of the remnant BH is defined by $a_{\mathrm{fl}} \equiv J_{\mathrm{BH,c}}/M_{\mathrm{BH,c}}^2$.

References

1. M. Shibata, Y. Sekiguchi, Phys. Rev. D **68**, 104020 (2003)
2. L. Baiotti, S. Bernuzzi, G. Corvino, R. de Pietri, A. Nagar, Phys. Rev. D **79**, 024002 (2009)
3. C. Reisswig, C.D. Ott, U. Sperhake, E. Schnetter, Phys. Rev. D **83**, 064008 (2011)
4. T. Regge, J.A. Wheeler, Phys. Rev. **108**, 1063 (1957)
5. F.J. Zerilli, Phys. Rev. D **2**, 2141 (1970)
6. V. Moncrief, Ann. Phys. (NY) **88**, 323 (1974)
7. E. Newman, R. Penrose, J. Math. Phys. **3**, 566 (1962)
8. E. Newman, R. Penrose, J. Math. Phys. **4**, 998 (1963)
9. R. Penrose, Phys. Rev. Lett. **10**, 66 (1963)
10. C. Reisswig, N.T. Bishop, D. Pollney, B. Szilágyi, Class. Quant. Grav. **27**, 075014 (2010)
11. H. Bondi, M.G.J. van der Burg, A.W.K. Metzner, Proc. Roy. Soc. Lond. **269**, 21 (1962)
12. R.K. Sachs, Proc. Roy. Soc. Lond. **270**, 103 (1962)
13. G.F.R. Ellis, Gen. Rel. Grav. **41**, 581 (2009)
14. C. Reisswig, D. Pollney, Class. Quant. Grav. **28**, 95015 (2011)
15. M. Shibata, K. Kyutoku, T. Yamamoto, K. Taniguchi, Phys. Rev. D **79**, 044030 (2009)
16. K. Kyutoku, M. Shibata, K. Taniguchi, Phys. Rev. D **82**, 044049 (2010)
17. K. Kyutoku, M. Shibata, K. Taniguchi, Phys. Rev. D **84**, 049902(E) (2011)
18. S.A. Teukolsky, Phys. Rev. Lett. **29**, 1114 (1972)
19. S.A. Teukolsky, Astrophys. J. **185**, 635 (1973)
20. P. Jaranowski, A. Królak, *Analysis of Gravitational-Wave Data* (Cambridge University Press, Cambridge, 2009)
21. M. Boyle, D.A. Brown, L.E. Kidder, A.H. Mroué, H.P. Pfeiffer, M.A. Scheel, G.B. Cook, S.A. Teukolsky, Phys. Rev. D **76**, 124038 (2007)
22. A. Buonanno, B.R. Iyer, E. Ochsner, Y. Pan, B.S. Sathyaprakash, Phys. Rev. D **80**, 084043 (2009)
23. M. Hannam, S. Husa, B. Brugmann, A. Gopakumar, Phys. Rev. D **78**, 104007 (2008)
24. M. Hannam, S. Husa, F. Ohme, D. Müller, B. Brügmann, Phys. Rev. D **82**, 124008 (2010)
25. L. Santamaria, F. Ohme, P. Ajith, B. Brügmann, N. Dorband, M. Hannam, S. Husa, P. Mösta, D. Pollney, C. Reisswig, E.L. Robinson, J. Seiler, B. Krishnan, Phys. Rev. D **82**, 064016 (2010)
26. L.E. Kidder, Phys. Rev. D **77**, 044016 (2008)
27. T. Damour, A. Nagar, Fundum. Theor. Phys. **162**, 211 (2011)
28. E. Berti, V. Cardoso, A.O. Starinets, Class. Quant. Grav. **26**, 163001 (2009)
29. B.D. Lackey, K. Kyutoku, M. Shibata, P.R. Brady, J.L. Friedman, Phys. Rev. D **85**, 044061 (2012)
30. T. Yamamoto, M. Shibata, K. Taniguchi, Phys. Rev. D **78**, 064054 (2008)
31. L. Smarr, Phys. Rev. D **7**, 289 (1973)

Chapter 6
The Merger of Nonspinning Black Hole–Neutron Star Binaries

We first study the nonspinning BH–NS binary merger to focus on the effect of the NS EOS, apart from the effect of the BH spin. The previous works by three groups [1–3] have found that the NSs in BH–NS binaries with high mass ratio $Q \gtrsim 4$ are barely subject to tidal disruption if the companion BH is not spinning. At the merger, the BH swallows most of the NS matter at one moment and the remnant disk mass is quite small or nearly equal to zero. Namely, the NS behaves approximately as a point particle even at the ISCO. Gravitational waves emitted in such a case have a similar waveform to that from a BH–BH binary. Because the behavior of NSs with high-mass BH companions does not show remarkable dependence on the EOS, they are unsuitable for the purpose of this chapter, i.e., to investigate the effect of the EOS on gravitational waves and final outcomes. Thus, we focus only on low mass-ratio binaries with $Q = 2$ and 3 in this chapter. Also, we choose relatively low-mass NSs, because two-piece EOSs adopted in this thesis may not be appropriate for modeling a high-mass NS with high central density, due to the lack of model parameters in the high-density region.

6.1 Models and Setup of AMR Grids

Table 6.1 summarizes key quantities for the initial models employed in the present numerical simulation. The labels for the models denote the name of the EOS, the mass ratio, and the NS mass; e.g., 2H-Q2M135 is modeled by 2H EOS, and its mass ratio and the NS mass are 2 and $1.35 M_{\odot}$, respectively. The primary purpose of this chapter is to study the dependence of gravitational waveforms and the final outcome on (i) the EOS of NSs, (ii) the mass ratio, and (iii) the NS mass. These purposes are reflected in our choice of the initial models.

We prepare quasiequilibrium states basically with the same value of $\Omega_0 m_0$ for the same value of Q irrespective of the EOS. The value of $\Omega_0 m_0$ is chosen to be small enough that the binaries spend more than 5 inspiral orbits before the onset of

Springer Theses, DOI: 10.1007/978-4-431-54201-8_6,

Table 6.1 Key parameters and quantities for the initial conditions adopted in the numerical simulations

Model	$G\Omega_0 m_0/c^3$	$M_*[M_\odot]$	$\mathscr{C}$	$\rho_{\max}(10^{14}\,\mathrm{g\,cm^{-3}})$	$M_0[M_\odot]$	$J_0[GM_\odot^2/c]$
2H-Q2M135	0.0250	1.455	0.1309	3.740	4.015	14.39
H-Q2M135	0.0280	1.484	0.1624	7.018	4.013	14.02
HB-Q2M135	0.0280	1.493	0.1718	8.262	4.013	14.02
HBs-Q2M135	0.0280	1.489	0.1723	9.154	4.013	14.02
HBss-Q2M135	0.0280	1.485	0.1741	1.082	4.013	14.02
B-Q2M135	0.0280	1.503	0.1819	9.761	4.013	14.02
Bs-Q2M135	0.0280	1.501	0.1856	1.137	4.013	14.02
Bss-Q2M135	0.0280	1.501	0.1940	1.490	4.013	14.02
2H-Q3M135	0.0280	1.455	0.1309	3.737	5.359	21.05
H-Q3M135	0.0300	1.484	0.1624	7.011	5.358	20.74
HB-Q3M135	0.0300	1.493	0.1718	8.254	5.358	20.74
B-Q3M135	0.0300	1.503	0.1819	9.751	5.357	20.74
2H-Q2M12	0.0220	1.282	0.1172	3.466	3.571	11.71
H-Q2M12	0.0280	1.303	0.1447	6.421	3.567	11.08
HB-Q2M12	0.0280	1.310	0.1527	7.522	3.567	11.08
B-Q2M12	0.0280	1.317	0.1614	8.832	3.567	11.08
HB-Q3M12	0.0280	1.310	0.1527	7.517	4.763	1.663
B-Q3M12	0.0280	1.317	0.1614	8.826	4.763	1.663

The initial angular velocity (Ω_0) in units of c^3/Gm_0, baryon rest mass (M_*), compactness of the NS in isolation ($\mathscr{C}$), maximum rest-mass density ($\rho_{\max}$), ADM mass of the system (M_0), and total angular momentum of the system (J_0), respectively. The labels for the models denote the name of the EOS, the mass ratio (Q), and the NS mass ($M_{\rm NS}$), where M135 (M12) stands for $M_{\rm NS} = 1.35M_\odot(1.2M_\odot)$

the merger. For $Q = 2$ binaries, a smaller value of initial angular velocity is required only for 2H EOS, because the NS with this EOS has a much larger radius than with other EOSs and is sensitive to the BH tidal force even for a larger orbital separation; to track $\gtrsim 5$ inspiral orbits before the tidal disruption, we have to choose the value of $\Omega_0 m_0$ by $\sim 10\,\%$ as small as that for other EOSs. For the case of $Q = 3$, we also choose smaller values of $\Omega_0 m_0$ for $M_{\rm NS} = 1.2M_\odot$ cases.

Table 6.2 summarizes the parameters of the grid structure for the simulations in this chapter. In all the simulations, we choose $(l_{\rm c}, l_{\rm f}) = (4, 4)$, and the value of L is chosen to be larger than the gravitational wavelengths at an initial instant $\lambda_0 \equiv \pi/\Omega_0$. Because the gravitational wavelength decreases during the evolution of the binaries, the outer boundary of the computational domains is guaranteed to be located in the wave zone throughout the simulation. In this chapter, we typically choose $N = 50$, with the exception that $N = 54$ for model Bss-Q2M135, in which the NS is quite compact and needs to be resolved with a better grid resolution. Each of the two finest domains covers the semimajor axis of the NS with 42–48 grid points and the BH radius (the coordinate radius of the AH) with typically $\approx$10 Q grid points, respectively. For several models arbitrarily chosen, we performed numerical simulations with lower grid resolutions, $N = 36$ and 42, to check the convergence of

Table 6.2 Setup of the grid structure for the computation with our AMR algorithm

Model	$\Delta x/M_0$	$R_{\rm diam}/\Delta x$	L/λ_0
2H-Q2M135	0.0471	90.8	2.377
H-Q2M135	0.0377	86.2	2.130
HB-Q2M135	0.0347	87.0	1.963
HBs-Q2M135	0.0353	85.2	1.996
HBss-Q2M135	0.0353	84.0	1.996
B-Q2M135	0.0330	85.1	1.863
Bs-Q2M135	0.0324	84.4	1.830
Bss-Q2M135	0.0270	95.4	1.650
2H-Q3M135	0.0353	89.0	1.996
H-Q3M135	0.0282	84.7	1.711
HB-Q3M135	0.0269	82.7	1.631
B-Q3M135	0.0247	83.8	1.497
2H-Q2M12	0.0565	86.9	2.510
H-Q2M12	0.0453	83.1	2.563
HB-Q2M12	0.0420	83.6	2.377
B-Q2M12	0.0392	83.4	2.218
HB-Q3M12	0.0306	84.6	1.713
B-Q3M12	0.0278	86.9	1.572

$\Delta x = h_7 = L/(2^7 N)$ is the grid spacing at the finest-resolution domain with L being the location of the outer boundaries for each axis. $R_{\rm diam}/\Delta x$ denotes the grid number assigned inside the semimajor diameter of the NS. λ_0 is the gravitational wavelength of the initial configuration. In all the models, both $l_{\rm c}$ and $l_{\rm f}$ are set to be 4

the numerical results (see the Appendix of [4, 5]). For $N = 54$ run, the total memory required for the simulations is about 11.6 Gbytes. We perform numerical simulations with personal computers of 12 Gbytes memory and of core-i7X processors with clock speed 3.2 or 3.33 GHz. We use 2–6 processors to perform one job with an OPEN-MP library. Typical computational time required to perform one simulation (for $\sim$40 ms in physical time of coalescence) is 4 weeks for 6 processor case.

6.2 Orbital Evolution and General Merger Process

To obtain a realistic numerical result for gravitational waveforms and the final outcome formed after the merger, it is necessary to exclude spurious effects associated with a noncircularity in the orbital motion as much as possible. To assess the circularity of the orbital motion, we plot the evolution of the coordinate separation $x^i_{\rm sep} = x^i_{\rm NS} - x^i_{\rm BH}$ for model HB-Q2M135 in Fig. 6.1. Here, the position of the maximum rest-mass density is identified as the coordinate of the NS, $x^i_{\rm NS}$, and the location of the puncture, $x^i_{\rm P}$, is the coordinate of the BH, $x^i_{\rm BH}$. This figure suggests that the orbital eccentricity appears to be low throughout the whole evolution. Because $\gtrsim$5 orbits are tracked, the eccentricity, which is likely to be nonzero initially, should

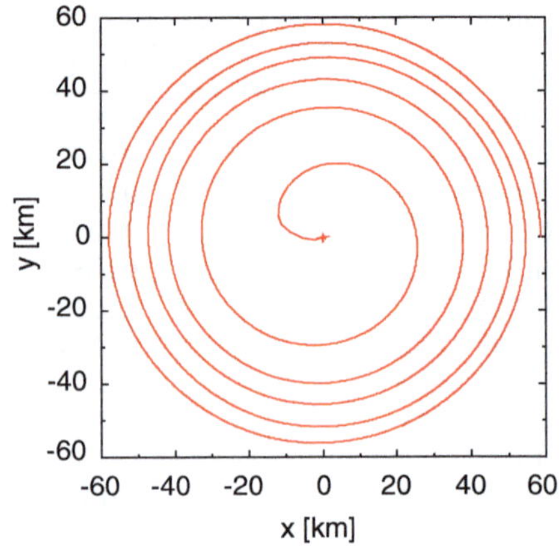

Fig. 6.1 Evolution of the coordinate separation of the binary $x^i_{\rm sep}$ for model HB-Q2M135

be suppressed by gravitational radiation reaction. We note that for all the models, similar trajectories are found.

The coordinate separation shown above is a gauge-dependent quantity. To show a stronger evidence that the eccentricity is suppressed to a small level, it is better to plot a gauge-independent quantity. Figure 6.2 plots the evolution of the orbital angular velocity defined from the $(l, m) = (2, 2)$ mode of Ψ_4 by

$$\Omega(t) = \frac{1}{2}\frac{|\Psi_4(l = m = 2)|}{|\int \Psi_4(l = m = 2)\mathrm{d}t|}, \tag{6.1}$$

for models 2H-Q2M135, H-Q2M135, HB-Q2M135, and B-Q2M135. Here, the horizontal axis is chosen to be an approximate retarded time defined by

$$t_{\rm ret} = t - D - 2M_0 \ln(D/M_0). \tag{6.2}$$

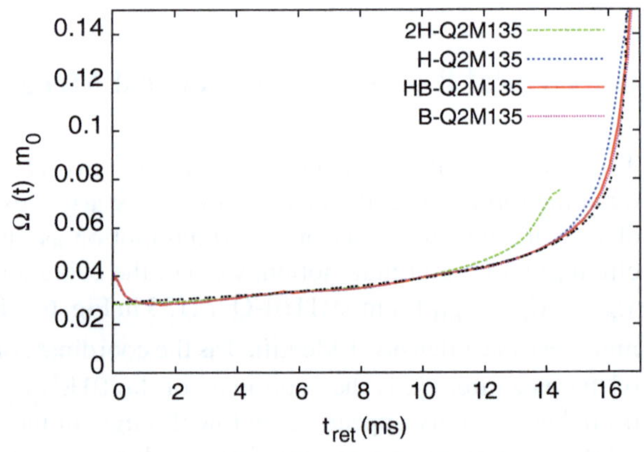

Fig. 6.2 Time evolution of the orbital angular velocity $\Omega(t)m_0$ for models 2H-Q2M135, H-Q2M135, HB-Q2M135, and B-Q2M135 as a function of a retarded time defined by (6.2) with an appropriate time shift. The dotted curve denotes the evolution of the orbital angular velocity calculated by the Taylor-T4 formula

We here do not plot the curve after the onset of tidal disruption. For comparison, the angular velocity derived from the Taylor-T4 formula is also plotted. To align the curve in the inspiral phase for $\Omega(t)m_0 \leq 0.05$, we appropriately shift the time for each model. For $t_{\rm ret} \gtrsim 0$ ms, an unphysical (a junk wave) component contained in the initial data dominates the waveform, and hence, $\Omega(t)$ derived from (6.1) does not give the angular velocity.

Figure 6.2 shows that the angular velocity obtained in numerical simulations agrees with that by the Taylor-T4 formula within a small modulation of $\Delta\Omega/\Omega \lesssim 5\,\%$ irrespective of the models. With the fact that the orbital eccentricity is approximately estimated as $e \approx \Delta\Omega/2\Omega$ for $e \ll 1$, we conclude that the orbital eccentricity is suppressed within $\sim$3 %. Figure 6.2 also shows that the deviation from the Taylor-T4 result becomes remarkable in an earlier time for models with stiffer EOSs such as 2H and H EOSs. This is due to the fact that the tidal elongation and disruption of the NS occur at slightly earlier stages of the inspiral orbits for models with the stiffer EOSs. This illustrates the fact that the stiffness of the EOS is reflected clearly in the gravitational-wave frequency (and gravitational-wave phase) as a function of time.

Figures 6.3 and 6.4 plot the snapshots of the rest-mass density profiles and the location of the AH on the equatorial plane at selected time slices for models 2H-Q2M12 and B-Q3M135. Figure 6.3 illustrates the process in which the NS is tidally disrupted to form a disk surrounding the companion BH. In this case, the NS is disrupted far outside the ISCO and then forms a one-armed spiral arm with large angular momentum. As a consequence of the angular momentum transport in the arm, a large amount of materials spread outward and then form a disk around the BH. We will report more details about the remnant disk in Sect. 6.5. Figure 6.4 illustrates the case in which the NS is not tidally disrupted before it is swallowed by the BH. In this case, the mass of the disk formed after the onset of the merger is negligibly small.

6.3 Gravitational Waveforms

Figures 6.5–6.8 plot the $(l, m) = (2, 2)$, plus-mode gravitational waveforms obtained numerically (hereafter referred to as h_+). All the waveforms are shown for an observer located along the z axis (axis perpendicular to the orbital plane) and plotted as a function of a retarded time $t_{\rm ret}$. We plot the amplitude in a normalized form, Dh_+/m_0, and the physical amplitude observed by an observer located at a hypothetical distance $D = 100\,$Mpc.

To validate the numerical waveforms, we compare them with the Taylor-T4 waveform, which is accurate up to 3.5PN order in phase and 3PN order in amplitude, with an appropriate time shift; the time shift is carried out to align the curve of $\Omega(t)$ as performed in Sect. 6.2. Figures 6.5–6.8 show that these two waveforms agree with each other irrespective of models during the inspiral phase, except for 2–3 initial cycles. The reasons for this initial disagreement are that an approaching

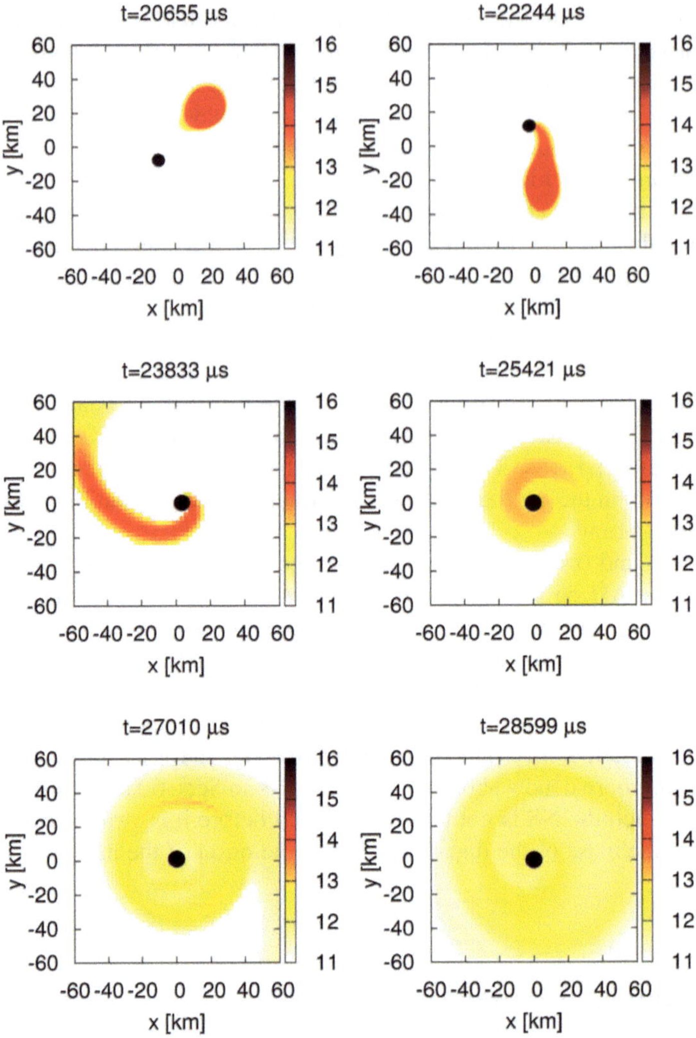

Fig. 6.3 Evolution of the rest-mass density profile in units of g cm^{-3} and the location of the AH on the equatorial plane for model 2H-Q2M135. The filled circles denote the regions inside the AHs. The color panels on the right-hand side of each figure show $\log_{10}(\rho)$

velocity associated with gravitational radiation reaction is not taken into account in the initial data and also the initial condition does not exactly model a quasicircular state, because we do not fully solve the Einstein equation for deriving it.

The numerical waveforms in the merger phase also (but due to a physical reason) deviate from the Taylor-T4 ones both in phase and amplitude, in particular for models with stiff EOSs, e.g., 2H-Q2M135 and 2H-Q2M12. For such models, ringdown waveforms associated with the BH quasinormal mode are not seen in the merger and

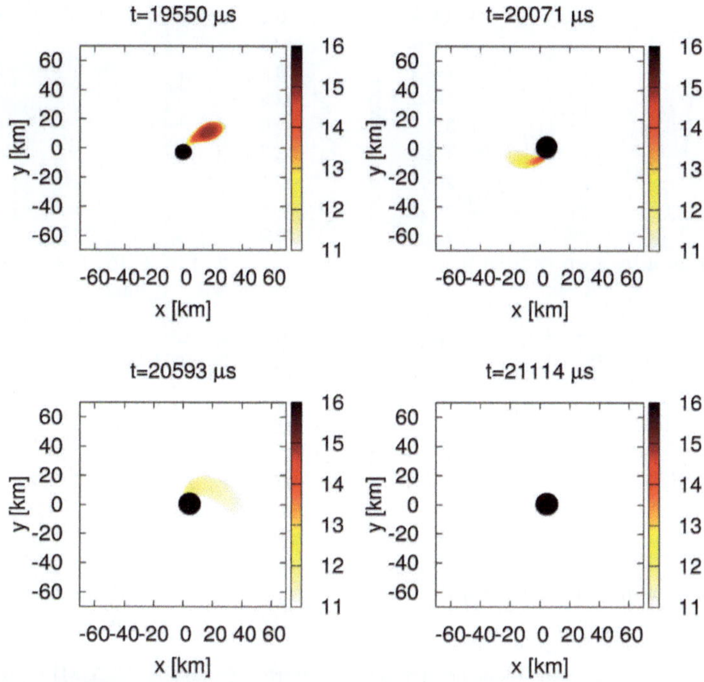

Fig. 6.4 The same as Fig. 6.3 but for model B-Q3M135

ringdown phases, and instead, the gravitational-wave amplitude damps suddenly in the middle of the inspiral phase. The reason for this quick damping is that the NS is tidally disrupted by the companion BH at an orbit in the inspiral phase within one orbital period, and then, the disrupted material forms a relatively low-density and nearly axisymmetric matter distribution around the BH, suppressing time variation of a mass quadrupole moment. Because the gravitational-wave emission stops in the middle of the inspiral motion, the maximum amplitude of gravitational waves is smaller for such a binary than for a binary with no tidal disruption, as shown in Figs. 6.5 and 6.6. All these facts illustrate that the finite size effect of the NS significantly modifies gravitational waves derived in the point-particle approximation (in the Taylor-T4 formula). On the other hand, ringdown gravitational waves are clearly seen for models with soft EOSs (for which tidal disruption does not occur) such as model B-Q3M135, in which the numerical and the Taylor-T4 waveforms are in more excellent agreement even in the late inspiral phase.

Table 6.3 presents total radiated energy ΔE and angular momentum ΔJ carried away by gravitational waves. The contribution from all the $l = 2$–4 modes is taken into account for ΔE and ΔJ. We estimate systematic errors in the presented values to be less than 10%, which are associated mainly with the finite grid resolution and partly with the finite extraction radii (cf. the Appendix of [4, 5]). We note that the $(l, |m|) = (2, 2)$ modes always contribute by $\gtrsim 90\,\%$ to both for ΔE and ΔJ.

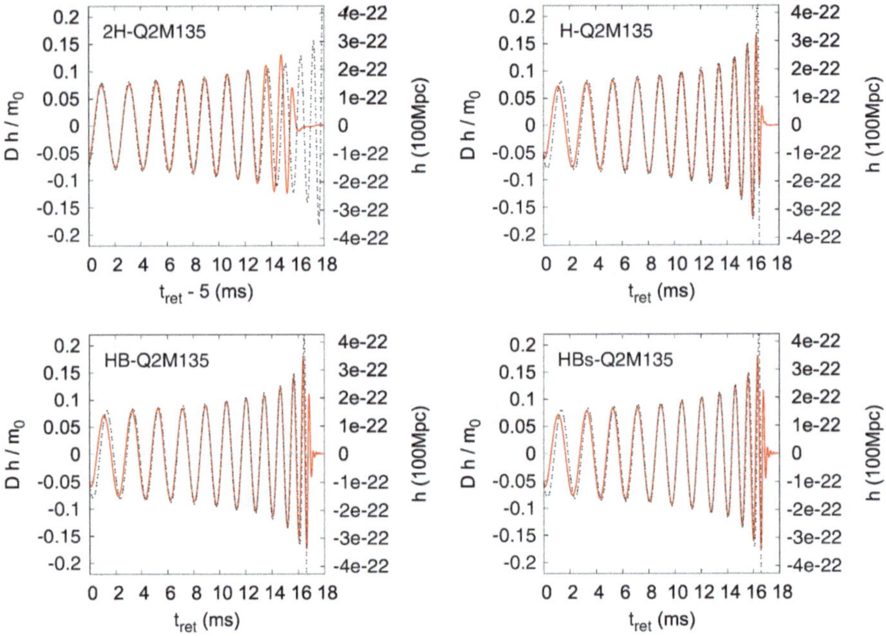

Fig. 6.5 $(l, m) = (2, 2)$, plus-mode gravitational waveforms for models 2H-Q2M135, H-Q2M135, HB-Q2M135, and HBs-Q2M135. All the waveforms are shown for an observer located along the z axis (axis perpendicular to the orbital plane) and plotted as a function of a retarded time. For model 2H-Q2M135, the waveform is plotted as a function of $t_{\rm ret} - 5\,{\rm ms}$ to align it with other waveforms (note that the initial value of Ω only for this model is smaller than those for other models). The left axis denotes the amplitude normalized by the distance from the binary D and the total mass m_0. The right axis denotes the physical amplitude of gravitational waves observed at a hypothetical distance 100 Mpc. The dotted curves denote the waveform calculated by the Taylor-T4 formula

The fraction of these modes is larger for binaries composed of less-compact NSs, because only binaries which escape the tidal disruption in the late inspiral phase can efficiently emit higher l-mode gravitational waves. Among other modes, (3, 3) and (4, 4) modes constitute most of the remaining part of ΔJ, whereas the order of magnitude of the (2, 1) mode is as large as that of the (4, 4) mode for ΔE.

The numerical results shown in Table 6.3 illustrate a quantitative dependence of gravitational-wave emission on the compactness of the NS: For a given mass ratio, gravitational-wave emission continues for a longer duration and consequently total radiated energy and angular momentum are larger for binaries composed of more compact NSs. Comparison among the models with $Q = 2$ and $M_{\rm NS} = 1.35 M_\odot$ and with the same initial value of Ωm_0 shows that both $\Delta E/M_0$ and $\Delta J/J_0$ are monotonically increasing functions of the NS compactness $\mathscr{C}$. This point is also recognized from Figs. 6.5–6.8, e.g., from the comparison among gravitational waves for models H-Q2M135, HB-Q2M135, and B-Q2M135 (note that for model 2H-Q2M135 the simulation is started from a lower value of Ωm_0 and it is not suitable

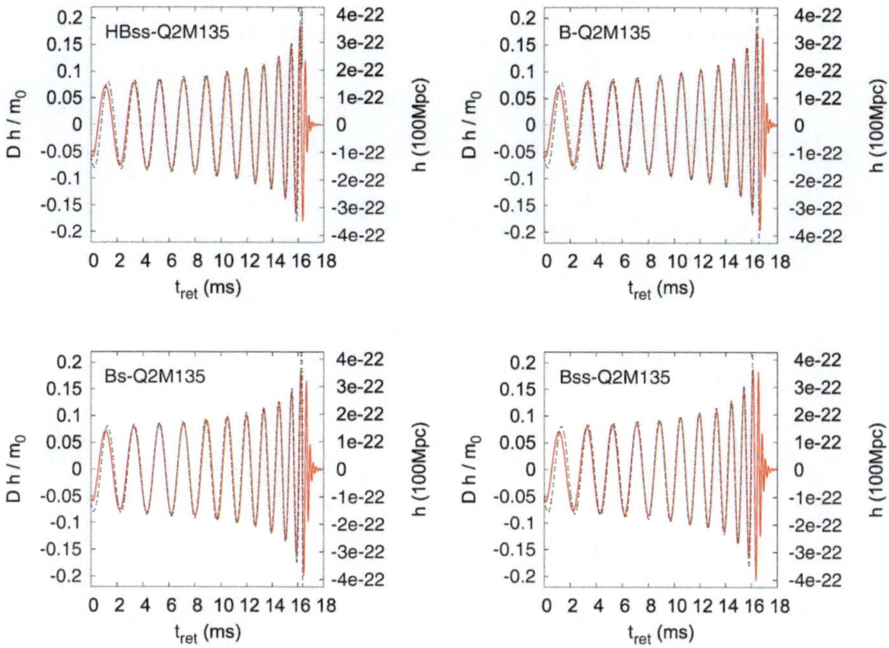

Fig. 6.6 The same as Fig. 6.5 but for models HBss-Q2M135, B-Q2M135, Bs-Q2M135, and Bss-Q2M135 is plotted as a function of $t_{\rm ret} - 9\,{\rm ms}$

for this comparison). Table 6.3 also shows that $\Delta J/\Delta E$ decreases as the EOS softens. This is due to the fact that $\Delta J/\Delta E \approx m/\Omega$ for a given angular harmonic of m, and for a soft EOS, more radiation is emitted at large angular velocity, Ω.

6.4 Gravitational-Wave Spectra

Characteristic features of a gravitational waveform, such as characteristic frequencies and their dependence on the EOS, are well reflected in the Fourier spectrum. Figures 6.9–6.11 display gravitational-wave spectra for all the models with the mass ratio $Q = 2$ and the models with the mass ratio $Q = 3$ and the NS mass $M_{\rm NS} = 1.35 M_\odot$. We define the Fourier spectrum as a sum of each Fourier component of two independent polarizations of the $(l, m) = (2, 2)$ mode as

$$\tilde{h}(f) \doteq \sqrt{\frac{|\tilde{h}_+(f)|^2 + |\tilde{h}_\times(f)|^2}{2}}, \tag{6.3}$$

$$\tilde{h}_A(f) = \int e^{2\pi i f t} h_A(t) {\rm d}t, \tag{6.4}$$

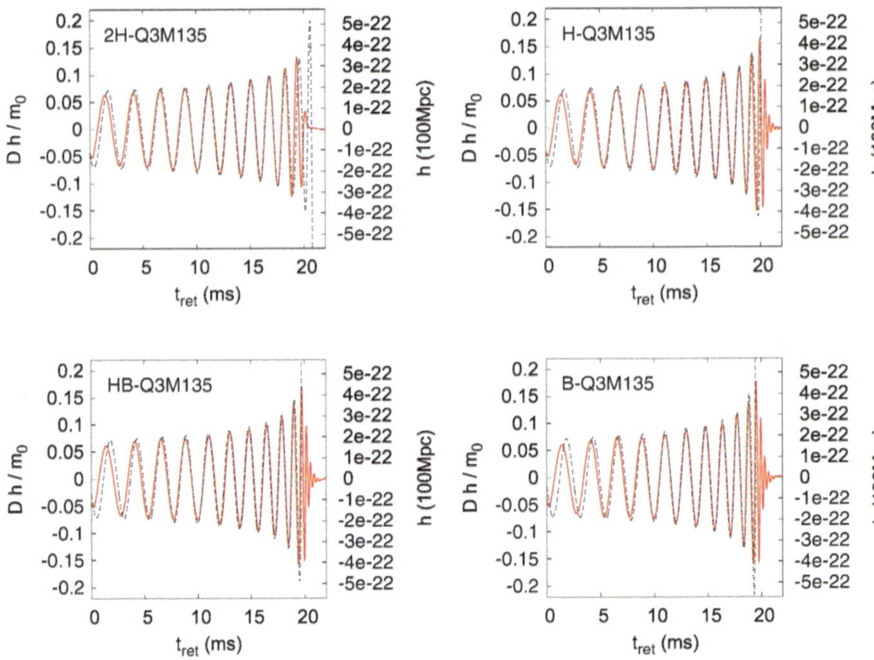

Fig. 6.7 The same as Fig. 6.5 but for models 2H-Q3M135, H-Q3M135, HB-Q3M135, and B-Q3M135

where A denotes two polarization modes, $+$ or $\times$. In calculating $\tilde{h}(f)$ from a numerically obtained Weyl scalar, Ψ_4, we always omit the unphysical radiation component extracted at $t_{\rm ret} \lesssim 0$ ms using a step function of retarded time as the window function so that the spurious radiation component does not introduce unphysical oscillations in the gravitational-wave spectrum. The spectrum amplitude for a low-frequency region of $f \approx \Omega(t_{\rm ret} = 0)/\pi$ changes slightly if we include the spurious radiation component. However, we believe that our use of the window function is physically reasonable. We always show the spectrum based on gravitational waves observed along the z axis (axis perpendicular to the orbital plane), which is the most optimistic direction for the gravitational-wave detection. (To obtain an averaged amplitude, we only need to multiply a factor of 0.4; e.g., see [8].) Because the Fourier components of any dimensionless quantity have the dimension of time, we define a dimensionless effective amplitude $f\tilde{h}(f)$. In the figure, we plot this quantity observed at a hypothetical distance 100 Mpc as a function of f(Hz) or a normalized amplitude $f\tilde{h}(f)D/m_0$ as a function of dimensionless frequency fm_0.

Figure 6.9 plots gravitational-wave spectra for $Q = 2$ and $M_{\rm NS} = 1.35M_\odot$ with all the EOSs employed in this chapter. For all these models, the total mass is universally $m_0 = 4.05M_\odot$, and thus, a nondimensional quantity, $fm_0(= Gfm_0/c^3)$, is plotted at the bottom and f in units of Hz is plotted at the top. Also, a normalized amplitude, $f\tilde{h}(f)D/m_0$, is plotted at the left side and $f\tilde{h}(f)$ observed at a distance

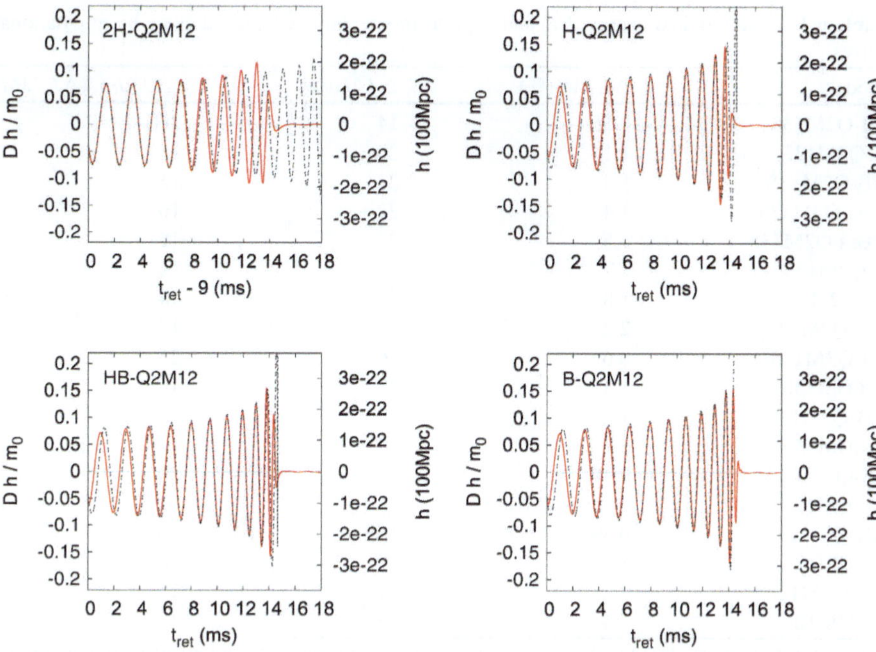

Fig. 6.8 The same as Fig. 6.5 but for models 2H-Q2M12, H-Q2M12, HB-Q2M12, and B-Q2M12. Again, the waveform for model 2H-Q2M12 is plotted as a function of $t_{\rm ret} - 9\,{\rm ms}$

of 100 Mpc from the binary is at the right side. For comparison, we also plot the spectra derived from the quadrupole formula (e.g., [9]) and the Taylor-T4 formula (dashed curves).

General qualitative features of the gravitational-wave spectrum by BH–NS binaries are summarized as follows. In the early stage of the inspiral phase, during which the orbital frequency is $\lesssim$1 kHz and the PN point-particle approximation works well, the gravitational-wave spectrum is approximately reproduced by the Taylor-T4 formula. For this phase, the spectrum amplitude of $f\tilde{h}(f)$ decreases as f^{-n_i} where $n_i = 1/6$ for $f \ll 1\,{\rm kHz}$ and the value of n_i increases with f for $f \lesssim 1\,{\rm kHz}$. As the orbital separation decreases, both the nonlinear effect of general relativity and the finite size effect of the NS come into play, and as a result, the PN point-particle approximation breaks down. If the tidal disruption sets in for a relatively large separation (e.g. for 2H EOS), the amplitude of the gravitational-wave spectra damps for a low frequency in the middle of the inspiral phase (before the ISCO is reached). By contrast, if the tidal disruption does not occur or occurs at a close orbit near the ISCO, the spectrum amplitude for a high frequency region ($f \gtrsim 1\,{\rm kHz}$) is larger than that predicted by the Taylor-T4 formula (i.e., the value of n_i decreases). In this case, an inspiral-like motion continues even inside the ISCO for a dynamical time scale and gravitational waves with a high amplitude are emitted. As a result, $f\tilde{h}(f)$

Table 6.3 Total radiated energy ΔE and angular momentum ΔJ carried away by gravitational waves

Model	$\Delta E/M_0(\%)$	$\Delta J/J_0(\%)$	$(\Delta J/J_0)/(\Delta E/M_0)$
2H-Q2M135	0.55	14	26
H-Q2M135	1.1	20	18
HB-Q2M135	1.4	22	16
HBs-Q2M135	1.4	22	16
HBss-Q2M135	1.5	23	15
B-Q2M135	1.7	24	14
Bs-Q2M135	1.8	25	14
Bss-Q2M135	2.2	27	12
2H-Q3M135	0.65	16	25
H-Q3M135	1.4	22	16
HB-Q3M135	1.6	23	15
B-Q3M135	1.7	24	14
2H-Q2M12	0.40	12	30
H-Q2M12	0.73	16	22
HB-Q2M12	0.89	18	20
B-Q2M12	1.1	20	18
HB-Q3M12	1.2	21	18
B-Q3M12	1.4	23	17

ΔE and ΔJ are normalized with respect to the initial ADM mass M_0 and angular momentum J_0, respectively. We also show the ratio between ΔJ and ΔE

becomes a slowly varying function of f for 1 kHz $\lesssim f \lesssim f_{\rm cut}$, where $f_{\rm cut} \sim$ 2–3 kHz is the so-called cutoff frequency which depends on the binary parameters as well as the EOS of the NSs. (A more strict definition of $f_{\rm cut}$ will be given below.) A steep damping of the spectra for $f \gtrsim f_{\rm cut}$ is universally observed, and for softer EOSs with a smaller radius of NSs, the frequency of $f_{\rm cut}$ is higher. This cutoff frequency is determined by the frequency of gravitational waves emitted when the NS is tidally disrupted for the stiff EOSs or by the frequency of a quasinormal mode of the formed BH for the soft EOSs. Therefore, the cutoff frequency provides potential information for the EOS through the tidal-disruption event of the NSs, in particular for the stiff EOSs.

Hereafter, we pay special attention to the cutoff frequency determined by the tidal disruption. It is natural to expect that the NS compactness $\mathscr{C}$ primarily determines the cutoff frequency in the combination, $f_{\rm cut}m_0$, because the orbital angular velocity at the onset of mass shedding, $R_{\rm shed}$, is written as a function of Q and $\mathscr{C}$ as [10, 11]

$$\Omega m_0 \propto \frac{\mathscr{C}^{3/2}(1+Q)^{3/2}}{\sqrt{Q}}. \tag{6.5}$$

In fact, a qualitative correlation between $\mathscr{C}$ and $f_{\rm cut}m_0$ was found in [1]. To reconfirm this, we first plot gravitational-wave spectra $f\tilde{h}(f)D/m_0$ as a function of fm_0] for $Q=2$ with the different NS mass $M_{\rm NS}=1.35M_\odot$ and $1.2M_\odot$ in Fig. 6.10. This

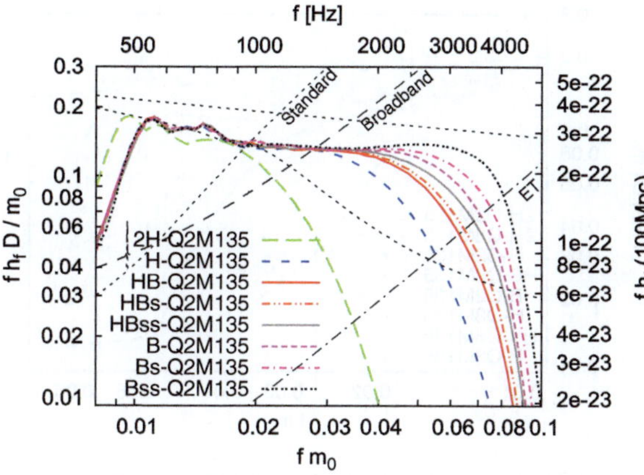

Fig. 6.9 Spectra of gravitational waves from BH-NS binaries for $Q = 2$ and $M_{\mathrm{NS}} = 1.35M_{\odot}$ with all the EOSs chosen in this chapter. The *bottom* axis denotes the normalized dimensionless frequency $fm_0 (= Gfm_0/c^3)$ and the *left* axis the normalized amplitude $f\tilde{h}(f)D/m_0$. The top axis denotes the physical frequency f in Hz and the *right* axis the effective amplitude $f\tilde{h}(f)$ observed at a distance of 100 Mpc from the binaries. The *short-dashed* slope line plotted in the *upper left* region denotes a planned noise curve of the Advanced-LIGO [6] optimized for $1.4M_{\odot}$ NS–NS inspiral detection ("Standard"), the *long-dashed* slope line denotes a noise curve optimized for the burst detection ("Broadband"), and the *dot-dashed* slope line plotted in the *lower right* region denotes a planned noise curve of the Einstein Telescope ("ET") [7]. The *upper* transverse *dashed line* is the spectrum derived by the quadrupole formula and the *lower* one is the spectrum derived by the Taylor-T4 formula, respectively

Fig. 6.10 The same as Fig. 6.9 but for $Q = 2$ and for $M_{\mathrm{NS}} = 1.35M_{\odot}$ and $1.2M_{\odot}$. Only the normalized amplitude $f\tilde{h}(f)D/m_0$ as a function of the dimensionless frequency fm_0 is shown

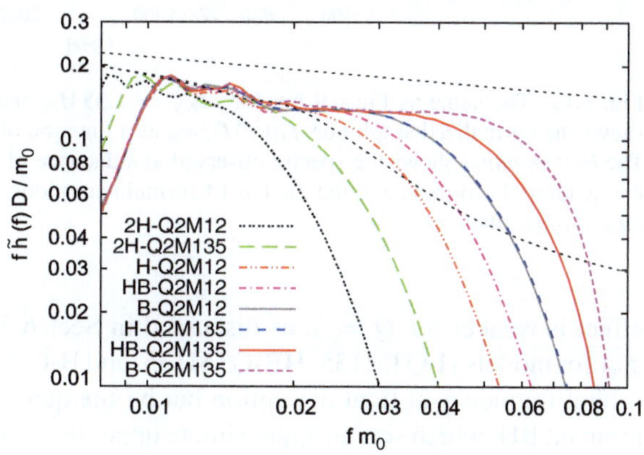

indeed shows $f_{\mathrm{cut}}m_0$ increases monotonically with $\mathscr{C}$ irrespective of the NS mass for the given mass ratio.

Figure 6.11 shows the gravitational-wave spectrum for $M_{\mathrm{NS}} = 1.35M_{\odot}$ and for $Q = 2$ and 3. The top panel plots $f\tilde{h}(f)D/m_0$ as a function of fm_0 and the bottom panel $f\tilde{h}(f)$ as a function of f for $D = 100$ Mpc. This shows that dependence of $f_{\mathrm{cut}}m_0$ on $\mathscr{C}$ for $Q = 3$ is weaker than for $Q = 2$. The reason for this is that the tidal

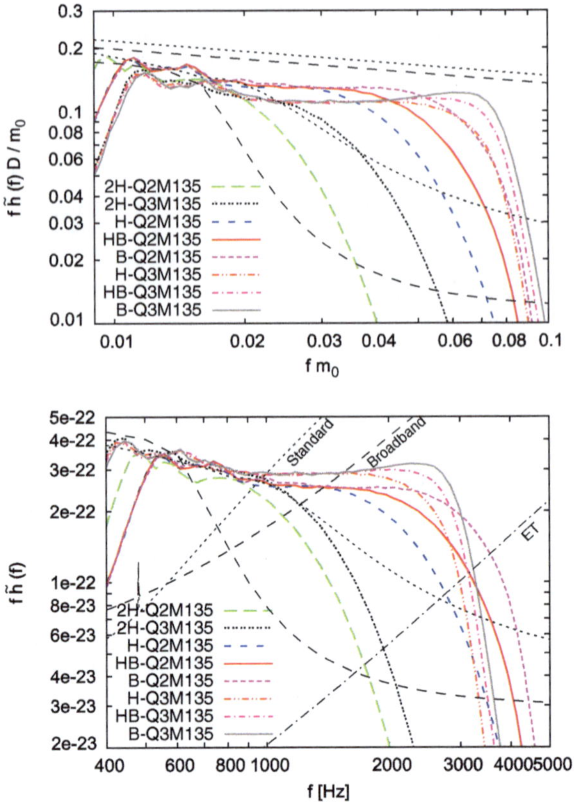

Fig. 6.11 The same as Fig. 6.9 but for $M_{\rm NS} = 1.35M_{\odot}$ and for $Q = 2$ and 3. The *top panel* shows the normalized amplitude $f\tilde{h}(f)D/m_0$ as a function of the dimensionless frequency fm_0. The *bottom panel* shows the spectra observed at a distance of 100 Mpc. The spectra derived from the quadrupole formula and the Taylor-T4 formula are plotted by the *short-dashed* ($Q = 2$) and *long-dashed* lines ($Q = 3$)

effect is weaker for $Q = 3$, as discussed in Sect. 6.5. (As later shown in Fig. 6.12, $f_{\rm cut}$ for models H-Q3M135, HB-Q3M135, and B-Q3M135 are not determined by the orbital frequency at tidal disruption but by the quasinormal-mode frequency of the remnant BH, which sets an approximate upper limit on the frequency of gravitational waves emitted in the merger.) Hence, the information of the EOS is not encoded in gravitational waves for $Q = 3$ as strongly as for $Q = 2$. The bottom panel shows that $f_{\rm cut}$ is between ~1 and 3 kHz depending weakly on the value of Q.

To analyze the cutoff frequency quantitatively and to strictly study its dependence on EOSs, we perform a systematic fitting procedure. As is done in [1], we fit all the spectra by a function with seven free parameters

$$\tilde{h}_{\rm fit}(f) = \tilde{h}_{\rm 3PN}(f)\mathrm{e}^{-(f/f_{\rm ins})^{\sigma_{\rm ins}}} + \frac{Am_0}{Df}\mathrm{e}^{-(f/f_{\rm cut})^{\sigma_{\rm cut}}}[1 - \mathrm{e}^{-(f/f_{\rm ins2})^{\sigma_{\rm ins2}}}], \quad (6.6)$$

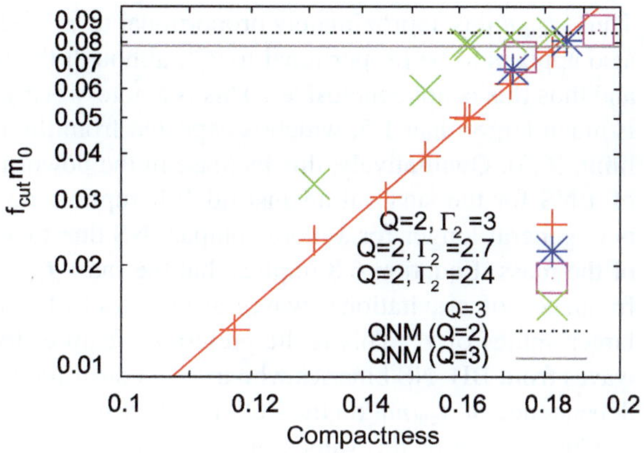

Fig. 6.12 $f_{\rm cut}m_0$ as a function of $\mathscr{C}$ in logarithmic scales. The solid line is obtained by a linear fitting of the data for $Q = 2$ and $\Gamma_2 = 3$. The short-dashed and long-dashed lines show approximate frequencies of quasinormal mode of the remnant BH for $Q = 2$ and $Q = 3$, respectively

where $\tilde{h}_{\rm 3PN}(f)$ is the Fourier spectrum calculated by the Taylor-T4 formula and $f_{\rm ins}$, $f_{\rm ins2}$, $f_{\rm cut}$, $\sigma_{\rm ins}$, $\sigma_{\rm ins2}$, $\sigma_{\rm cut}$, and A are free parameters. The first and second terms of (6.6) denote the spectrum models for the inspiral and merger phases, respectively. We determine these free parameters by searching the minimum for a weighted norm defined by

$$\sum_i \left\{ [f_i\tilde{h}(f_i) - f_i\tilde{h}_{\rm fit}(f_i)] f_i^{1/3} \right\}^2 , \tag{6.7}$$

where i denotes the data point for the spectrum. In [1], $\sigma_{\rm ins} = 3.5$ and $\sigma_{\rm ins2} = 5$ are fixed to save the computational costs. Here, these are chosen to be free parameters to reproduce a more consistent spectrum with the original one.

Among these seven free parameters, we focus on $f_{\rm cut}$ because it depends most strongly on the compactness $\mathscr{C}$ and the EOS of the NS. Figure 6.12 plots $f_{\rm cut}m_0$, obtained in this fitting procedure, as a function of $\mathscr{C}$. Also the typical quasinormal-mode frequencies, $f_{\rm QNM}$, of the remnant BH calculated in Sect. 6.5 are plotted by the two horizontal lines, which show that the values of $f_{\rm cut}m_0$ for models H-Q3M135, HB-Q3M135, and B-Q3M135 agree approximately with $f_{\rm QNM}$ and indicates that $f_{\rm cut}$ for these models are irrelevant to the tidal disruption. For $Q = 3$, $f_{\rm cut}m_0$ depends clearly on the EOS only for $\mathscr{C} \lesssim 0.16$. This agrees with the result with $\Gamma = 2$ polytropic EOS [1]. By contrast, $f_{\rm cut}m_0$ for $Q = 2$ depends strongly on the NS compactness $\mathscr{C}$ irrespective of $M_{\rm NS}$ not only for the piecewise polytropic EOS but also for $\Gamma = 2$ polytrope [1]. The solid line in Fig. 6.12 is the linear fitting of $\ln(f_{\rm cut}m_0)$ as a function of $\ln(\mathscr{C})$ for $Q = 2$ and for the piecewise polytrope with $\Gamma_2 = 3$, and denoted by a well-approximated relation

$$\ln(f_{\rm cut}m_0) = (3.87 \pm 0.12) \ln \mathscr{C} + (4.03 \pm 0.22). \tag{6.8}$$

Thus, $f_{\rm cut}m_0$ is approximately proportional to $\mathscr{C}^{3.9}$ (for $Q = 3$ and $\Gamma_2 = 3$, $f_{\rm cut}m_0$ also appears to be proportional to $\mathscr{C}^4$, although the number of data points is small and thus this is not conclusive). This is a note-worthy point because the power of $\mathscr{C}$ is much larger than 1.5, which is expected from the relation for the mass-shedding limit, (6.5). Qualitatively, this increase in the power is natural because the duration of a NS for the survival against tidal disruption after the onset of mass shedding is in general longer for a more compact NS due to a stronger central condensation of the mass. Equation 6.8 implies that the ratio $f_{\rm cut}/f_{\rm shed}$ (>1), where $f_{\rm shed}$ is the frequency of gravitational waves at the onset of mass shedding, is larger for the larger values of $\mathscr{C}$. This is the preferable feature, for an observer of gravitational waves from BH–NS binaries who tries to constrain the EOS of the NSs, because the dependence of $f_{\rm cut}m_0$ on the EOS is enhanced.

Comparison of the values of $f_{\rm cut}m_0$ for models HB-Q2M135 ($\Gamma_2 = 3.0$ and $\mathscr{C} = 0.1718$), HBs-Q2M135 ($\Gamma_2 = 2.7$ and $\mathscr{C} = 0.1723$), and HBss-Q2M135 ($\Gamma_2 = 2.4$ and $\mathscr{C} = 0.1741$), for which the value of $\mathscr{C}$ is approximately identical, shows that $f_{\rm cut}m_0$ depends also on the adiabatic index of EOS in the central region, Γ_2. The reason for this is that the NSs with smaller values of Γ_2 (but with the same value of $\mathscr{C}$) have more centrally condensed density profile as can be seen from the value of $\rho_{\rm max}$ in Table 6.1, and hence, are less subject to tidal disruption ($f_{\rm cut}m_0$ becomes larger). Quantitatively, the value of $f_{\rm cut}m_0$ increases by ∼20 %, when the value of Γ_2 is varied from 3 to 2.4. This result suggests that it may be possible to constrain not only the compactness of a NS but also its density profile, e.g., the tidal deformability of the NS [12], and detailed function of $P(\rho)$ for the EOS, if gravitational waves emitted during the merger of low-mass BH–NS binaries are detected.

6.5 Properties of the Disk

If a NS is tidally disrupted before it is swallowed by the companion BH, a disk may be formed around the BH. Figure 6.13 plots the time evolution of the rest mass of the material located outside the AH $M_{r>r_{\rm AH}}$ defined by (5.31). This shows that most of the material is swallowed by the BH soon after the onset of the merger (or tidal disruption) within ∼1 ms, but 1–10 % of total rest mass survives around the BH to be a disk, if the tidal disruption occurs (see Table 6.4 which lists the numerical results of $M_{r>r_{\rm AH}}$ at $t - t_{\rm merger} \approx 10$ ms for all the models).

To clarify that the disk will survive for a time duration longer than the dynamical time scale of the system, we estimate an accretion time scale. Figure 6.13 shows that for $t - t_{\rm merger} \gtrsim 5$ ms, $M_{r>r_{\rm AH}}$ for each model behaves approximately as $C\exp(-t/t_{\rm d})$ where C is a constant and $t_{\rm d}$ is the accretion time scale which we determine by a least-square fitting of $M_{r>r_{\rm AH}}(t)$ at $t - t_{\rm merger} \approx 10$ ms. The fourth column of Table 6.4 lists the numerical results. It is found that the accretion time scale is always longer than

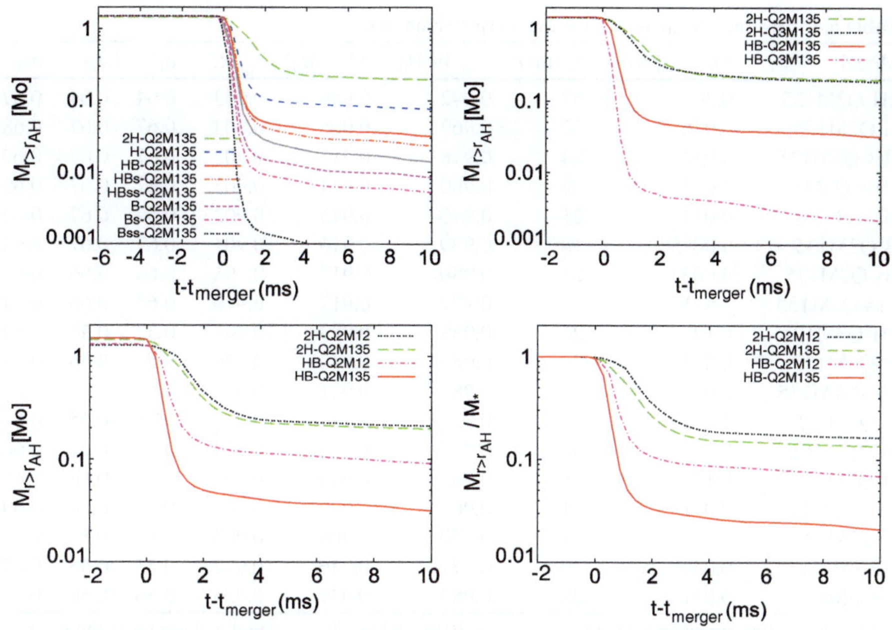

Fig. 6.13 Evolution of the rest mass of the material located outside the AH, $M_{r>r_{\rm AH}}$, with an appropriate time shift; in these plots, the time at the onset of the merger is taken as the time origin. The *top-left* panel shows the results for models with $Q = 2$ and $M_{\rm NS} = 1.35M_\odot$ for all the EOSs employed in this chapter. The *top-right* panel shows the results for selected models with $M_{\rm NS} = 1.35M_\odot$ but with different values of Q. The *bottom-left* panel shows the results for selected models with $Q = 2$ but with the different NS mass $M_{\rm NS}$. The *bottom-right* panel is the same as the *bottom-left* panel except for the normalization of the mass, with respect to the initial rest mass M_*

the dynamical time scale of the remnant disk ~10 ms, and hence, we conclude that the BH–NS merger always forms a long-lived accretion disk, if the disk is formed.[1]

Figure 6.14 plots the values of $M_{r>r_{\rm AH}}$ estimated at $t - t_{\rm merger} \approx 10$ ms as a function of the NS compactness $\mathscr{C}$ and clarifies the dependence of the disk mass on the EOS. This figure summarizes the key features as follows: (i) for a given mass ratio and for a given adiabatic index of the core, Γ_2, the disk mass decreases monotonically with the increase of $\mathscr{C}$ for $M_{r>r_{\rm AH}} \lesssim 0.1M_\odot$; (ii) for a given mass ratio and for a given NS compactness, the disk mass increases slightly with the increase of Γ_2; and (iii) the disk mass is highly sensitive to the mass ratio of the binary, Q, for a given mass and EOS of the NS. In the following, we observe these features from Fig. 6.13 in detail.

The top left panel of Fig. 6.13 plots the disk-mass evolution for binaries with $Q = 2$, $M_{\rm NS} = 1.35M_\odot$ and for all the EOSs employed in this chapter. For this sample, $\mathscr{C} \propto R_{\rm NS}^{-1}$ since $M_{\rm NS}$ is identical, and we find that the disk mass increases

[1] Note that in the presence of magnetic fields, angular momentum transport by them works efficiently, and thus, the accretion time scale may be shorter than that presented here in reality.

Table 6.4 Several key quantities for the merger remnants

Model	$M_{r>r_{\rm AH}}[M_\odot]$	$t_{\rm d}$ (ms)	$C_{\rm e}/4\pi M_0$	$M_{\rm irr}/M_0$	$C_{\rm p}/C_{\rm e}$	$a_{\rm f}$	$a_{\rm f2}$	$a_{\rm f1}$
2H-Q2M135	0.20	57	0.942	0.886	0.913	0.64	0.64	0.67
H-Q2M135	0.076	32	0.969	0.905	0.903	0.67	0.67	0.68
HB-Q2M135	0.032	24	0.978	0.912	0.902	0.67	0.67	0.69
HBs-Q2M135	0.024	22	0.980	0.914	0.902	0.67	0.67	0.69
HBss-Q2M135	0.014	21	0.980	0.915	0.902	0.67	0.67	0.69
B-Q2M135	0.0085	18	0.980	0.916	0.904	0.67	0.67	0.68
Bs-Q2M135	0.0053	23	0.980	0.917	0.906	0.66	0.66	0.68
Bss-Q2M135	7×10^{-4}	...	0.977	0.917	0.910	0.65	0.65	0.67
2H-Q3M135	0.19	26	0.958	0.923	0.945	0.52	0.52	0.54
H-Q3M135	0.013	26	0.982	0.940	0.936	0.56	0.56	0.58
HB-Q3M135	0.0022	25	0.983	0.941	0.936	0.56	0.55	0.57
B-Q3M135	2×10^{-4}	...	0.982	0.941	0.938	0.55	0.55	0.57
2H-Q2M12	0.21	66	0.937	0.885	0.918	0.62	0.62	0.66
H-Q2M12	0.12	28	0.958	0.900	0.907	0.66	0.66	0.68
HB-Q2M12	0.091	31	0.965	0.902	0.905	0.66	0.66	0.69
B-Q2M12	0.065	27	0.970	0.906	0.903	0.67	0.67	0.69
HB-Q3M12	0.044	30	0.977	0.936	0.937	0.55	0.55	0.57
B-Q3M12	0.011	28	0.982	0.939	0.935	0.56	0.56	0.58

All the quantities are estimated at $t - t_{\rm merger} \approx 10\,$ms, where $t_{\rm merger}$ denotes the time of the merger. $M_{r>r_{\rm AH}}$ is the rest mass of the disk surrounding the BH; because the accretion is still ongoing at the end of simulations due to the hydrodynamic angular momentum transport process, the values listed give only an approximate mass of the long-lived accretion disk, which survives for a time scale longer than the dynamical time scale $\sim$10 ms. $t_{\rm d}$ is the approximate accretion time scale estimated around $\approx$10 ms after the merger, which we show only for the case $M_{r>r_{\rm AH}} \gtrsim 0.001 M_\odot$. $C_{\rm e}$ and $C_{\rm p}$ are the circumferential radii of the AH along the equatorial plane and meridional plane, respectively, and $C_{\rm e}/4\pi$ is the approximate mass of the remnant BH. $M_{\rm irr}$ is the irreducible mass of the remnant BH. $a_{\rm f}$ is the nondimensional spin parameter of the remnant BH estimated from $C_{\rm p}/C_{\rm e}$. $a_{\rm f2}$ and $a_{\rm f1}$ are also the nondimensional spin parameters, estimated from the quantities on the AH and approximate conservation laws, respectively

monotonically with $\mathscr{C}^{-1}$ (see Table 2.1 for $\mathscr{C}$ of each model); the disk mass is larger for a model for which the tidal disruption occurs at a more distant orbit (i.e., for a smaller value of $f_{\rm cut}$, cf. Fig. 6.12). This is quite reasonable because the earlier onset of tidal disruption helps more materials to remain outside the ISCO of the BH.

Comparison of the results for models HB-Q2M135 ($\Gamma_2 = 3.0$ and $\mathscr{C} = 0.1718$), HBs-Q2M135 ($\Gamma_2 = 2.7$ and $\mathscr{C} = 0.1723$), and HBss-Q2M135 ($\Gamma_2 = 2.4$ and $\mathscr{C} = 0.1741$) indicates that the disk mass depends not only on the compactness of the NS but also on the adiabatic index of the core, Γ_2; a higher value of Γ_2 is preferable for forming a massive disk. This dependence on Γ_2 is consistent with the result reported in [13]; the NS with a larger value of the adiabatic index is more subject to tidal disruption (tidal disruption occurs for more distant orbital separation). The physical interpretation for this result is that the degree of central mass concentration for NSs of larger values of the adiabatic index is weaker, helping earlier tidal disruption (in

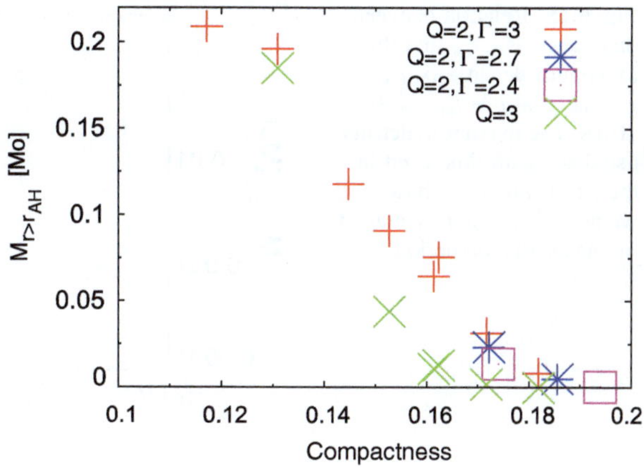

Fig. 6.14 Disk mass $M_{r>r_{\rm AH}}$ at $t - t_{\rm merger} \approx 10\,{\rm ms}$ as a function of the NS compactness $\mathscr{C}$

other words, we may say that the tidal Love number or deformability is larger for the larger value of Γ_2).

The top right panel of Fig. 6.13 plots the disk-mass evolution for the NS with the same mass ($M_{\rm NS} = 1.35 M_\odot$) but with different mass ratio $Q = 2$ and $Q = 3$ and with HB and 2H EOSs. This, together with Fig. 6.14, shows that the disk mass depends strongly on the mass ratio, in particular for the soft EOS. The reason for this is simply that the NS is less subject to tidal disruption for a larger BH mass (i.e., for weaker tidal force near the ISCO). The present result suggests that the disk mass is much smaller than $0.01 M_\odot$ for BH–NS binaries with the typical NS mass of $M_{\rm NS} = 1.2$–$1.35 M_\odot$ and $\mathscr{C} \gtrsim 0.16$, if the BH is nonspinning and $M_{\rm BH} \gtrsim 4 M_\odot$. Only for the case $\mathscr{C} \lesssim 0.16$, the disk mass may be larger than $0.01 M_\odot$ even with a high-mass BH companion. This conclusion is in agreement with the previous studies [1–3].

The two bottom panels of Fig. 6.13 compare the disk-mass evolution for models 2H-Q2M12 and 2H-Q2M135 and for models HB-Q2M12 and HB-Q2M135. In the left panel we plot the disk mass in units of $M_\odot$ while the bottom right panel plots the disk mass in units of M_*. We note that the NS radius depends weakly on the mass for $1.2 M_\odot \le M_{\rm NS} \le 1.35 M_\odot$ for both EOSs, and also the mass ratio Q is identical for these models. Nevertheless, the disk mass depends strongly on the NS mass except for models with stiff 2H EOS as seen in Table 6.4; it decreases with the increase of $M_{\rm NS}$. Thus, not the NS radius $R_{\rm NS}$ but $\mathscr{C}$ is the key parameter for determining the disk mass.

Before closing this section, we summarize several key properties of the remnant disk. Figure 6.15 plots the relation between $M_{r>r_{\rm AH}}$ and the maximal rest-mass density $\rho_{\rm max}$ of the remnant disk estimated at $t - t_{\rm merger} \approx 10$ ms. This clearly shows a strong correlation between two quantities. The value of $M_{r>r_{\rm AH}}$ increases approximately linearly with $\rho_{\rm max}$ for $M_{r>r_{\rm AH}} \lesssim 0.1 M_\odot$, and for $M_{r>r_{\rm AH}} \ge 0.01 M_\odot$, $\rho_{\rm max}$ is larger than $4 \times 10^{11}\,{\rm g\,cm^{-3}}$. Because the density is high and the temperature should be also high enough ($\sim$10 MeV if viscous effects or magnetohydrodynamic

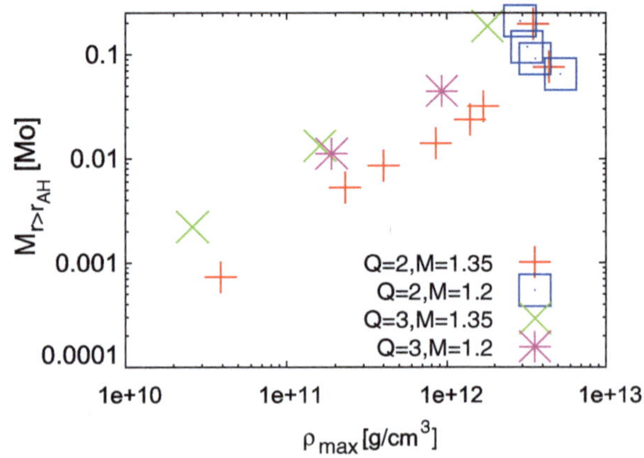

Fig. 6.15 Relation between disk mass $M_{r>r_{\rm AH}}$ and the maximum density, $\rho_{\rm max}$, estimated at $t - t_{\rm merger} \approx 10$ ms. The maximum density oscillates with time even in the quasistationary phase, and we here plot a value averaged in one oscillation period

effects are taken into account [14–16]), neutrinos will be copiously produced in such a disk in reality. Because of the high density and temperature, the cross section to the nucleon will be large enough ($\sim 10^{-41}$ cm^2) to trap neutrinos inside the disk of nucleon number density $n_{\rm n} = \rho/m_{\rm n} \gtrsim 10^{35}\,{\rm cm}^{-3}$ where $m_{\rm n} = 1.66 \times 10^{-24}$ g is nucleon mass [17–19]. Therefore, a neutrino-dominated accretion disk will be always produced, if BH–NS binaries result in a system composed of the BH and surrounding disk of mass larger than $0.01 M_\odot$.

6.6 Properties of the Remnant BH

Table 6.4 shows several quantities associated with the remnant BH such as the mass and spin, in addition to the disk mass. Unlike the disk mass, the mass and spin of the remnant BH depend weakly on the EOS of the NS. For given values of Q and $M_{\rm NS}$, the BH mass tends to be slightly smaller for stiffer EOS, primarily because the fraction of the NS mass swallowed by the BH is smaller (the disk mass is larger). The spin does not show such a clear dependence. The reason is that the spin angular momentum of the remnant BH is affected by two competing processes; one is the orbital angular momentum dissipation due to gravitational radiation reaction and the other is the distribution of the angular momentum to the disk surrounding the BH. The former dissipation effect is important for the case in which the NS is compact and the tidal disruption does not occur as stated in Sect. 6.3. By contrast, the latter effect is more important for the case in which the NS is less compact and the tidal disruption occurs in the relatively early stage of the inspiral phase. Although the relation $\Delta J > J_{r>r_{\rm AH}}$ always holds for all the models, we may also have the relation $\Delta E \gtrsim M_{r>r_{\rm AH}}$. As a result, the nondimensional spin parameter of the remnant BH depend very weakly on the EOS. For comparison, $a_{\rm f1}$ and $a_{\rm f2}$ defined in Chap. 5 are also shown in Table 6.4. It is found that $a_{\rm f}$ and $a_{\rm f2}$ agree with each other within the error of $\Delta a = 0.001$. By contrast, a_{f1} does not agree well with the other two estimates, as is found in [1], particularly when the massive remnant disk is formed: the maximum error is $\Delta a \approx 0.04$. Taking into account the fact that the agreement between $C_e/4\pi$ and $M_{BH,f}$ is always better than 0.5 %, a possible reason for this discrepancy is that (5.32) systematically underestimates the angular momentum of the disk. Hereafter, we only refer to a_f as the nondimensional spin parameter of the remnant BH.

The spin of the remnant BH is primarily determined by the mass ratio, Q; $a_f = 0.66 \pm 0.03$ for $Q = 2$ and $a_f = 0.54 \pm 0.02$ for $Q = 3$ (here $\pm$ signs do not imply the error bars but signify differences due to the EOS). Thus, the spin parameter is modified by the EOS only in ± 5 %.

From the typical value of the spin parameter a_f and mass of the remnant BH $M_{BH,f}$, we estimate typical quasinormal-mode frequencies f_{QNM} of the remnant BH by the fitting formula [20]

$$f_{QNM} M_{BH,f} \approx \frac{1}{2\pi} \left[1.5251 - 1.1568(1 - a_f)^{0.1292} \right]. \tag{6.9}$$

Then, $f_{QNM} \approx 0.083/M_{BH,f}$ for $Q = 2$ and $\approx 0.076/M_{BH,f}$ for $Q = 3$, respectively. Assuming that $C_e/4\pi$ gives an approximate value of $M_{BH,f}$, these values are in good agreement with the ringdown part of gravitational waves for models Bss-Q2M135 and B-Q3M135, for which the disk masses are negligibly small, respectively. We note that this estimation is valid only when the quasinormal modes of the BH are excited, and actually the tidal disruption of the NS often suppresses the quasinormal-mode excitation as can be seen in Figs. 6.5–6.8, in particular, for the stiff EOS such as 2H.

References

1. M. Shibata, K. Kyutoku, T. Yamamoto, K. Taniguchi, Phys. Rev. D **79**, 044030 (2009)
2. Z.B. Etienne, Y.T. Liu, S.L. Shapiro, T.W. Baumgarte, Phys. Rev. D **79**, 044024 (2009)
3. M.D. Duez, F. Foucart, L.E. Kidder, H.P. Pfeiffer, M.A. Scheel, S.A. Teukolsky, Phys. Rev. D **78**, 104015 (2008)
4. K. Kyutoku, M. Shibata, K. Taniguchi, Phys. Rev. D **82**, 044049 (2010)
5. K. Kyutoku, M. Shibata, K. Taniguchi, Phys. Rev. D **84**, 049902(E) (2011)
6. B. Abbott et al., Astrophys. J. **683**, L45 (2009)
7. M. Punturo et al., Class. Quant. Grav. **27**, 194002 (2010)
8. A. Buonanno, G.B. Cook, F. Pretorius, Phys. Rev. D **75**, 124018 (2007)
9. C. Cutler, É.É. Flanagan, Phys. Rev. D **49**, 2658 (1994)
10. K. Taniguchi, T.W. Baumgarte, J.A. Faber, S.L. Shapiro, Phys. Rev. D **75**, 084005 (2007)
11. K. Taniguchi, T.W. Baumgarte, J.A. Faber, S.L. Shapiro, Phys. Rev. D **77**, 044003 (2008)
12. B.D. Lackey, K. Kyutoku, M. Shibata, P.R. Brady, J.L. Friedman, Phys. Rev. D **85**, 044061 (2012)
13. M. Ishii, M. Shibata, Y. Mino, Phys. Rev. D **71**, 044017 (2005)
14. S. Setiawan, M. Ruffert, H.T. Janka, Astron. Astrophys. **458**, 553 (2006)
15. W.H. Lee, E. Ramirez-Ruiz, D. Page, Astrophys. J. **632**, 421 (2005)
16. M. Shibata, Y.I. Sekiguchi, R. Takahashi, Prog. Theor. Phys. **118**, 257 (2007)

17. R. Narayan, T. Piran, P. Kumar, Astrophys. J. **557**, 949 (2001)
18. T. Di Matteo, R. Perna, R. Narayan, Astrophys. J. **579**, 706 (2002)
19. K. Kohri, S. Mineshige, Astrophys. J. **577**, 311 (2002)
20. E. Berti, V. Cardoso, A.O. Starinets, Class. Quant. Grav. **26**, 63001 (2009)

Chapter 7
The Merger of Spinning Black Hole–Neutron Star Binaries

In this chapter, numerical simulations are performed for a wide range of nondimensional BH spin parameter, a, as well as for a variety of the mass ratios, Q. For nonspinning BH–NS binaries, we already found that the low mass ratio of $Q \lesssim 3$ is required for tidal disruption of NSs to occur sufficiently outside the ISCO of the BH unless the EOS is extremely stiff. If the tidal disruption occurs inside or at an orbit very close to the ISCO, we do not see strong effects of the tidal disruption. In such cases, gravitational waveforms are similar to those of a BH–BH binary even in the merger phase, and the mass of the remnant disk is negligible [1]. However, the allowed range of the mass ratio for the tidal disruption is modified drastically for a BH–NS binary with the prograde BH spin [2, 3] because the ISCO radius[1] of the BH with a prograde spin becomes smaller by a factor of 1–6 [4] than that of the nonspinning BH with the same mass. Strong spin effects for the tidal disruption are also found in the numerical-relativity simulation of the spinning BH–NS binary merger with a simplified, Γ-law EOS [5]. In this chapter, we perform a more systematic study of the tidal disruption for different EOSs, masses of each component, and BH spins.

7.1 Models and Setup of AMR Grids

Tables 7.1 and 7.2 summarize several key quantities for the initial conditions in our numerical simulations. The labels for the model denote the EOS name, the mass ratio, the NS mass, and the nondimensional spin parameter of the BH. Specifically, "a75," "a5," and "a-5" correspond to the spin parameters $a = 0.75, 0.5$, and -0.5, respectively. For example, HB-Q3M135a5 means that the EOS is HB and $(Q, M_{\rm NS}, a) = (3, 1.35 M_\odot, 0.5)$. Although we vary the NS mass systematically, the results of the merger remnant are reported only for binaries with $M_{\rm NS} = 1.35 M_\odot$

[1] In this thesis, "the ISCO radius" always represents "the ISCO radius in the Boyer-Lindquist coordinates," as is described in Sect. 1.3.

K. Kyutoku, *The Black Hole–Neutron Star Binary Merger in Full General Relativity*, Springer Theses, DOI: 10.1007/978-4-431-54201-8_7,

Table 7.1 Key parameters and quantities for the initial conditions adopted in numerical simulations for $Q = 2$ and 3

Model	$G\Omega_0 m_0/c^3$	$M_*[M_\odot]$	$\mathscr{C}$	$\rho_{max}(10^{14}\,\mathrm{g\,cm^{-3}})$	$M_0[M_\odot]$	$J_0[GM_\odot^2/c]$
2H-Q2M135a75	0.025	1.455	0.1309	3.740	4.014	13.83
1.5H-Q2M135a75	0.028	1.468	0.1456	5.104	4.012	13.42
H-Q2M135a75	0.028	1.484	0.1624	7.019	4.012	13.42
HB-Q2M135a75	0.028	1.493	0.1718	8.263	4.012	13.42
B-Q2M135a75	0.028	1.503	0.1819	9.762	4.012	13.42
2H-Q2M135a5	0.025	1.455	0.1309	3.740	4.014	14.02
1.5H-Q2M135a5	0.028	1.468	0.1456	5.104	4.012	13.63
H-Q2M135a5	0.028	1.484	0.1624	7.018	4.012	13.63
HB-Q2M135a5	0.028	1.493	0.1718	8.263	4.012	13.63
B-Q2M135a5	0.028	1.503	0.1819	9.762	4.012	13.63
2H-Q2M135a-5	0.022	1.455	0.1309	3.740	4.019	15.15
H-Q2M135a-5	0.025	1.484	0.1624	7.018	4.017	14.74
HB-Q2M135a-5	0.028	1.493	0.1718	8.262	4.015	14.41
B-Q2M135a-5	0.028	1.503	0.1819	9.760	4.015	14.41
2H-Q2M12a75	0.025	1.282	0.1172	3.465	3.568	10.93
H-Q2M12a75	0.028	1.303	0.1447	6.421	3.566	10.60
HB-Q2M12a75	0.028	1.310	0.1527	7.523	3.566	10.60
B-Q2M12a75	0.028	1.317	0.1614	8.833	3.566	10.60
2H-Q2M145a75	0.025	1.572	0.1401	3.926	4.312	15.96
H-Q2M145a75	0.028	1.607	0.1744	7.452	4.309	15.48
HB-Q2M145a75	0.028	1.617	0.1848	8.811	4.309	15.48
B-Q2M145a75	0.028	1.629	0.1960	10.46	4.309	15.48
2H-Q3M135a75	0.028	1.455	0.1309	3.737	5.357	20.00
1.5H-Q3M135a75	0.030	1.468	0.1456	5.100	5.355	19.64
H-Q3M135a75	0.030	1.484	0.1624	7.013	5.355	19.64
HB-Q3M135a75	0.030	1.493	0.1718	8.256	5.355	19.64
B-Q3M135a75	0.030	1.503	0.1819	9.753	5.355	19.63
2H-Q3M135a5	0.028	1.455	0.1309	3.737	5.357	20.36
1.5H-Q3M135a5	0.030	1.468	0.1456	5.100	5.356	20.02
H-Q3M135a5	0.030	1.484	0.1624	7.012	5.356	20.01
HB-Q3M135a5	0.030	1.493	0.1718	8.255	5.356	20.01
B-Q3M135a5	0.030	1.503	0.1819	9.753	5.356	20.01
HB-Q3M135a-5	0.030	1.493	0.1718	8.253	5.359	21.46
2H-Q3M145a75	0.028	1.572	0.1401	3.923	5.754	23.07
H-Q3M145a75	0.030	1.607	0.1744	7.445	5.751	22.65
HB-Q3M145a75	0.030	1.617	0.1848	8.803	5.751	22.65
B-Q3M145a75	0.030	1.629	0.1960	10.45	5.751	22.65

The initial angular velocity (Ω_0) in units of c^3/Gm_0, baryon rest mass (M_*), compactness of the NS in isolation ($\mathscr{C}$), maximum rest-mass density (ρ_{max}), ADM mass of the system (M_0), and total angular momentum of the system (J_0), respectively. The labels for the models denote the name of the EOS, the mass ratio (Q), the NS mass (M_{NS}), and the nondimensional spin parameter of the BH (a), where M135 (M12, M145) stands for $M_{NS} = 1.35M_\odot(1.2M_\odot, 1.45M_\odot)$, and "a75," "a5," and "a-5" correspond to the spin parameters $a = 0.75, 0.5$, and -0.5, respectively. (See Table 6.1 in Chap. 6 for models of nonspinning BH-NS binaries)

Table 7.2 The same as Table 7.1 but for $Q = 4$ and 5

Model	$G\Omega_0 m_0/c^3$	$M_*[M_\odot]$	$\mathscr{C}$	$\rho_{\max}$ (10^{14} g cm^{-3})	$M_0[M_\odot]$	$J_0[GM_\odot^2/c]$
2H-Q4M135a75	0.030	1.455	0.1309	3.735	6.702	26.07
H-Q4M135a75	0.032	1.484	0.1624	7.007	6.700	25.62
HB-Q4M135a75	0.032	1.493	0.1718	8.249	6.700	25.63
B-Q4M135a75	0.032	1.503	0.1819	9.746	6.700	25.62
2H-Q4M135a5	0.035	1.455	0.1309	3.732	6.698	25.64
H-Q4M135a5	0.035	1.484	0.1624	7.004	6.698	25.63
HB-Q4M135a5	0.035	1.493	0.1718	8.244	6.698	25.63
B-Q4M135a5	0.035	1.503	0.1819	9.740	6.698	25.63
2H-Q5M135a75	0.036	1.455	0.1309	3.730	8.044	30.95
H-Q5M135a75	0.036	1.484	0.1624	7.000	8.044	30.95
HB-Q5M135a75	0.036	1.493	0.1718	8.241	8.044	30.95
B-Q5M135a75	0.036	1.503	0.1819	9.736	8.043	30.95

in this chapter because the difference in the NS mass complicates the properties of the remnant, such as the mass of the disk. Results for $M_{\rm NS} \neq 1.35 M_\odot$ are analyzed only for gravitational waves.

For the same value of the mass ratio, we basically prepare the initial conditions with the same value of the initial angular velocity, Ω_0, normalized by the total mass of the binary, $\Omega_0 m_0$. For 2H EOS, in which the NS radius is the largest, we exceptionally adopt a smaller value of $\Omega_0 m_0$ than for other EOSs to guarantee $\gtrsim$5 orbits before tidal disruption occurs. The reason for this is that the tidal disruption occurs for a large orbital separation in 2H EOS. When the BH has a prograde spin, the number of orbits to the merger for a given value of $\Omega_0 m_0$ increases due to spin-orbit repulsive interaction [6], compared to the nonspinning BH case. On the other hand, when the BH has a retrograde spin, the number of orbits decreases due to spin-orbit attractive interaction. For $a = -0.5$, the number of orbits is typically by ~ 1 orbit smaller than for $a = 0$. For this reason, we also prepare the initial condition with a smaller value of $\Omega_0 m_0$ for H EOS and $a = -0.5$.

Tables 7.3 and 7.4 summarize the parameters of the grid structure for our simulations. The structure of the AMR grids depends primarily on the mass ratio of the binary because the distances between two objects and the center of mass depend strongly on the mass ratio for our initial models. Specifically, we choose $(l_{\rm c}, l_{\rm f}) = (4, 4)$ for all binaries with $M_{\rm NS} = 1.35 M_\odot$ and $Q = 2, 3,$ and 4. We choose $(l_{\rm c}, l_{\rm f}) = (3, 5)$ for binaries with $Q = 5$. For binaries with $M_{\rm NS} \neq 1.35 M_\odot$, we choose $(l_{\rm c}, l_{\rm f}) = (3, 4)$ because we do not evaluate disk masses for them. In all the simulations, L is chosen to be larger than or comparable to the gravitational wavelengths at an initial instant $\lambda_0 \equiv \pi/\Omega_0$. We always choose $N = 50$ for the best resolved runs in the work of this chapter. One of the two finest regions covers the semimajor axis of the NS by ~42–45 grid points. The other covers the coordinate radius of the AH typically by ~5–10Q grid points, depending on the BH spin. We also perform simulations with $N = 36$ and 42 for several arbitrary chosen models to

Table 7.3 Setup of the grid structure for the simulation with our AMR algorithm for $Q = 2$ and 3

Model	l_c	l_f	$\Delta x/M_0$	$R_{diam}/\Delta x$	L/λ_0
2H-Q2M135a75	4	4	0.0471	90.8	2.386
1.5H-Q2M135a75	4	4	0.0426	87.7	2.417
H-Q2M135a75	4	4	0.0377	86.2	2.138
HB-Q2M135a75	4	4	0.0347	87.1	1.968
B-Q2M135a75	4	4	0.0324	86.7	1.837
2H-Q2M135a5	4	4	0.0471	90.8	2.378
1.5H-Q2M135a5	4	4	0.0426	87.7	2.410
H-Q2M135a5	4	4	0.0377	86.2	2.131
HB-Q2M135a5	4	4	0.0347	87.2	1.962
B-Q2M135a5	4	4	0.0324	86.7	1.831
2H-Q2M135a-5	4	4	0.0470	90.7	2.092
H-Q2M135a-5	4	4	0.0376	86.4	1.902
HB-Q2M135a-5	4	4	0.0347	87.1	1.962
B-Q2M135a-5	4	4	0.0324	86.7	1.831
2H-Q2M12a75	3	4	0.0583	84.7	1.476
H-Q2M12a75	3	4	0.0442	85.3	1.252
HB-Q2M12a75	3	4	0.0410	85.7	1.162
B-Q2M12a75	3	4	0.0389	84.2	1.102
2H-Q2M145a75	3	4	0.0461	85.2	1.166
H-Q2M145a75	3	4	0.0347	85.3	0.985
HB-Q2M145a75	3	4	0.0316	87.1	0.896
B-Q2M145a75	3	4	0.0292	87.1	0.829
2H-Q3M135a75	4	4	0.0367	85.5	2.084
1.5H-Q3M135a75	4	4	0.0326	84.0	1.986
H-Q3M135a75	4	4	0.0282	84.7	1.718
HB-Q3M135a75	4	4	0.0260	85.6	1.581
B-Q3M135a75	4	4	0.0235	87.9	1.431
2H-Q3M135a5	4	4	0.0353	88.9	1.997
1.5H-Q3M135a5	4	4	0.0326	84.0	1.980
H-Q3M135a5	4	4	0.0282	84.7	1.712
HB-Q3M135a5	4	4	0.0260	85.7	1.576
B-Q3M135a5	4	4	0.0243	85.3	1.471
HB-Q3M135a-5	4	4	0.0260	85.7	1.576
2H-Q3M145a75	3	4	0.0328	87.7	0.933
H-Q3M145a75	3	4	0.0250	87.4	0.760
HB-Q3M145a75	3	4	0.0234	86.6	0.712
B-Q3M145a75	3	4	0.0214	87.7	0.651

l_c and l_f are the number of coarser domains and a half of finer domains, respectively. $\Delta x = h_l = L/(2^l N)$ $(l = l_c + l_f - 1)$ is the grid spacing at the finest-resolution domain with L being the location of the outer boundaries along each axis. $R_{diam}/\Delta x$ denotes the grid number assigned inside the semimajor diameter of the NS. λ_0 is the gravitational wavelength of the initial configuration. (See Table 6.2 in Chap. 6 for models with nonspinning BHs)

Table 7.4 The same as Table 7.3 but for $Q = 4$ and 5

Model	l_c	l_f	$\Delta x/M_0$	$R_{diam}/\Delta x$	L/λ_0
2H-Q4M135a75	4	4	0.0296	83.4	1.804
H-Q4M135a75	4	4	0.0223	84.5	1.450
HB-Q4M135a75	4	4	0.0203	86.5	1.319
B-Q4M135a75	4	4	0.0190	85.8	1.237
2H-Q4M135a5	4	4	0.0296	83.2	2.097
H-Q4M135a5	4	4	0.0219	85.9	1.548
HB-Q4M135a5	4	4	0.0205	85.5	1.448
B-Q4M135a5	4	4	0.0188	86.4	1.332
2H-Q5M135a75	3	5	0.0235	86.3	1.718
H-Q5M135a75	3	5	0.0180	86.2	1.314
HB-Q5M135a75	3	5	0.0167	86.3	1.224
B-Q5M135a75	3	5	0.0159	84.6	1.159

check the convergence of the results and find approximately the same level of convergence as that found in the merger of nonspinning BH–NS binaries (see the Appendix of [7–9]). For the $Q = 5$ runs, the total memory required is about 11.8 Gbytes. We perform numerical simulations with personal computers of 12 Gbytes memory and of core-i7X processors with clock speeds of 3.2 or 3.33 GHz. We use 2–6 processors to perform one job with an OPEN-MP library. The typical computational time required to perform one simulation (for ~50 ms in physical time of coalescence for the $a = 0.75$ case) is 4 weeks for the 6 processor case.

7.2 Overview of the Merger Process

Figure 7.1 plots the evolution of the coordinate separation defined by $x^i_{sep} = x^i_{NS} - x^i_{BH}$ for models HB-Q2M135a5, HB-Q2M135, and HB-Q2M135a-5, for which $\Omega_0 m_0$ takes the same values. Here, x^i_{NS} is the position of the maximum rest-mass density and x^i_{BH} is the location of the puncture, x^i_P. Figure 7.1 shows that the numbers of orbits increases as the BH spin increases from retrograde to prograde [5]. Specifically, the number of orbit are ~7, 5.5, and 4 for $a = 0.5, 0$, and -0.5, respectively. This difference comes primarily from the spin-orbit interaction between these two angular momenta [10]; in the PN approximation, a force proportional to the inner product of the orbital and spin angular momenta of two objects appears at 1.5PN order. Here, we do not have to consider the NS spin angular momentum in the assumption of the irrotational velocity field and, therefore, we only consider the interaction between the orbital and BH spin angular momenta throughout this thesis. When these two angular momenta are parallel and the inner product is positive ($a > 0$), an additional repulsive force works between the BH and NS. This repulsive force reduces the orbital angular velocity because the centrifugal force associated

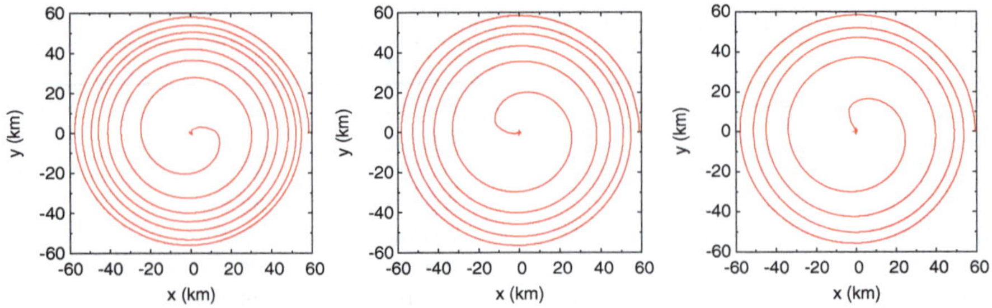

Fig. 7.1 Evolution of the orbital separation $x^i_{\rm sep} = x^i_{\rm NS} - x^i_{\rm BH}$ of binaries with $(Q, M_{\rm NS}) = (2, 1.35M_\odot)$ and HB EOS. The *left*, *middle*, and *right panels* show the results with the prograde BH spin $a = 0.5$, zero BH spin $a = 0$, and retrograde BH spin $a = -0.5$, respectively

with the orbital motion can be reduced, and hence, the luminosity of gravitational radiation, which is proportional to $\Omega^{10/3}$, is also reduced. This strong dependence of the luminosity on Ω makes the approaching velocity smaller in the late inspiral phase, and, therefore, the number of orbits increases. Conversely, when these two angular momenta are antiparallel ($a < 0$), an additional attractive force increases the angular velocity and gravitational-wave luminosity in the late inspiral phase. In this case, the orbital separation decreases faster due to a larger approaching velocity, and the number of orbits becomes smaller as the retrograde BH spin increases. All these results agree qualitatively with those of [5].

The fate of BH–NS binaries is classified into two categories. One is the case in which the NS is disrupted by the BH tidal field before the BH swallows the NS, and the other is the case in which the BH swallows the NS without tidal disruption. In this chapter, we focus mainly on the former case. We plot snapshots of the rest-mass density profiles and the location of the AH on the equatorial plane at selected time slices for models HB-Q3M135a75, HB-Q3M135a5, and HB-Q3M135a-5 in Figs. 7.2, 7.3, 7.4, respectively. The NS is disrupted outside the ISCO in the $a > 0$ cases (Figs. 7.2 and 7.3) and forms a one-armed spiral structure with a large angular momentum. The material in the inner part of the spiral arm gradually falls onto the BH due to angular momentum transport via hydrodynamic torque in the spiral arm. The material with a sufficiently large specific angular momentum escapes the capture by the BH and forms an accretion disk, which survives for a time much longer than the dynamical time scale $\sim$ a few ms. We note that the prompt infall of the one-armed spiral structure onto the BH occurs from a relatively narrower region for $a = 0.5$ than for $a = 0.75$. The reason is that the inner edge of the spiral arm contacts the AH well before the arm becomes nearly axisymmetric due to a large radius of the AH and ISCO for $a = 0.5$. The infall of the disrupted material from a narrow region of the BH frequently occurs when the NS is tidally disrupted in a binary with a *high mass ratio*, whereas this is rare in a binary with a nonspinning BH because the NS is not disrupted in a high mass-ratio binary. This difference in the merger process is well-reflected in gravitational waveforms (see Sect. 7.6). By contrast, the NS is

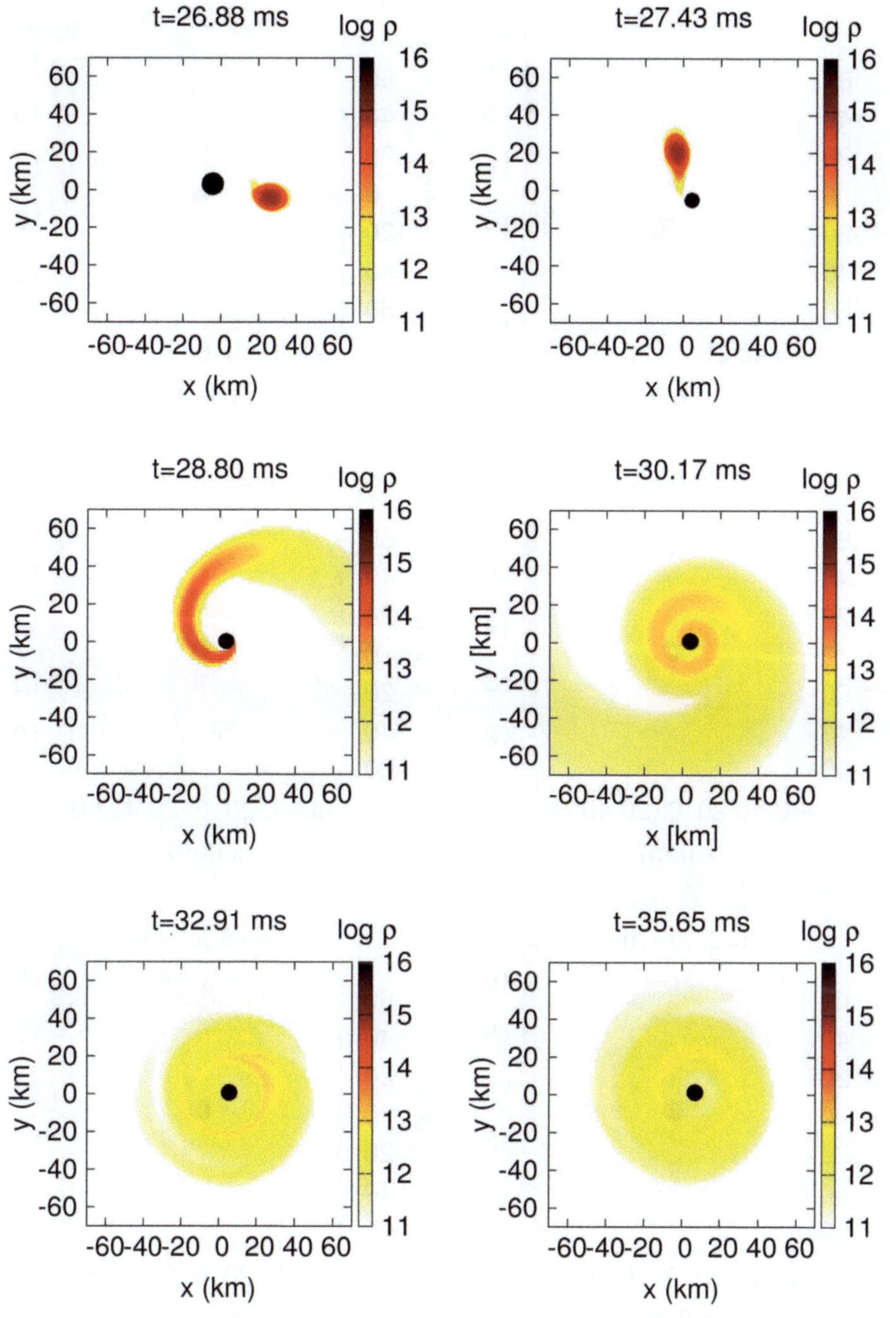

Fig. 7.2 Evolution of the rest-mass density profile in units of $\mathrm{g\,cm^{-3}}$ and the location of the AH on the equatorial plane for model HB-Q3M135a75. The *filled circle* denotes the region inside the AH. The color panel on the right of each plot show $\log_{10}(\rho)$

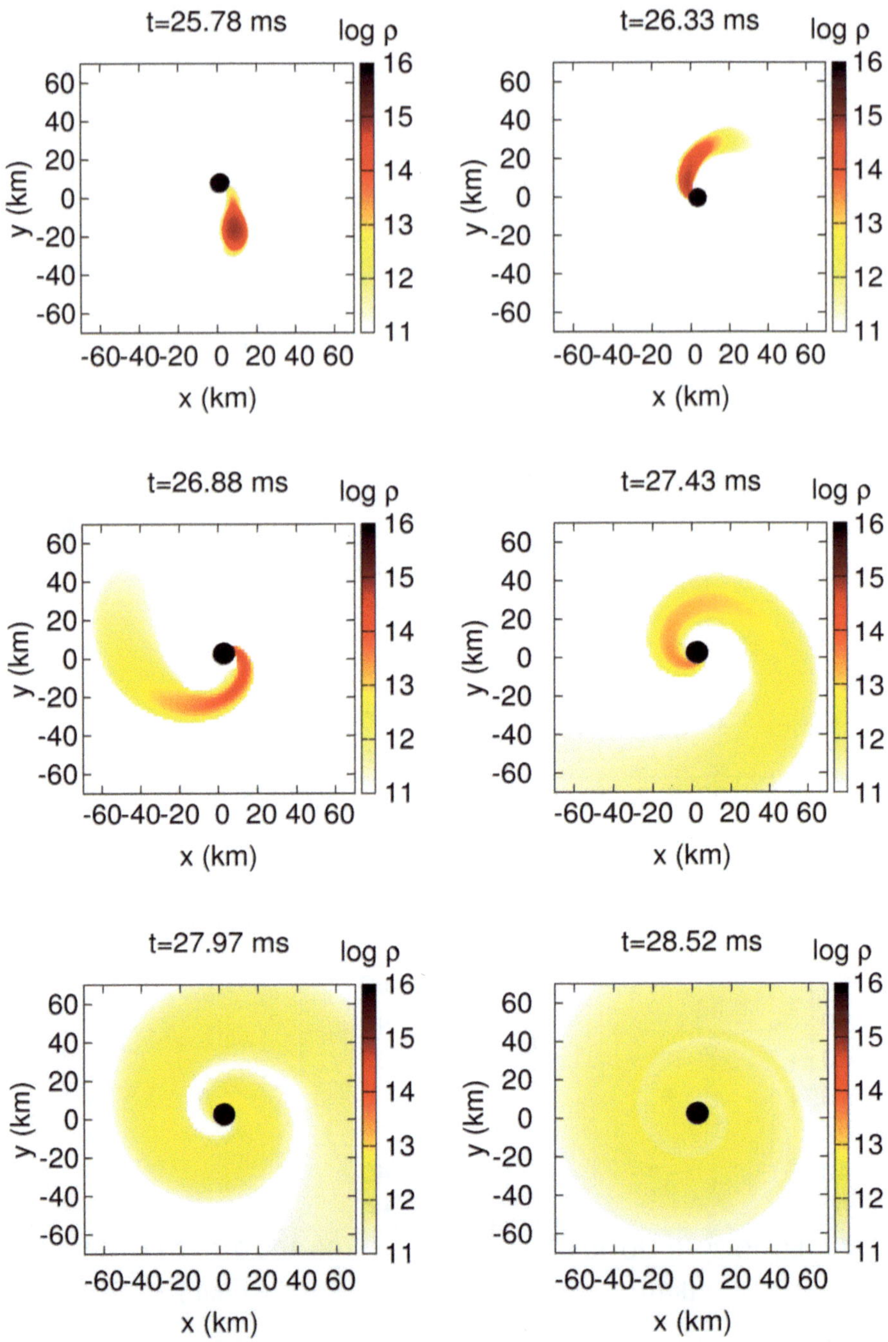

Fig. 7.3 The same as Fig. 7.2 but for model HB-Q3M135a5

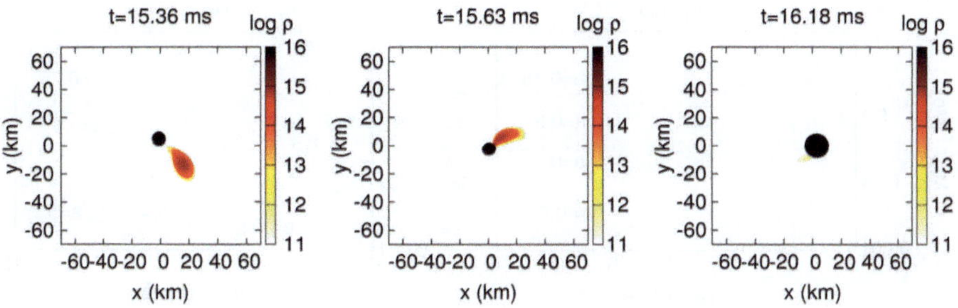

Fig. 7.4 The same as Fig. 7.2 but for model HB-Q3M135a-5

swallowed by the BH without tidal disruption, and essentially no material is left outside the ISCO for model HB-Q3M135a-5 (Fig. 7.4).

Note that the feature of the NS tidal disruption appears very weakly not only for model HB-Q3M135a-5 but also for model HB-Q3M135 ($a = 0$) because the mass ratio $Q = 3$ is so high that the tidal effect is less important for the nonspinning BH with the typical NS radius ~11–12 km. The enhancement of the tidal effect by a prograde BH spin results primarily from the decrease of the BH ISCO radius [4]. In the Boyer-Lindquist coordinates, a Kerr BH has an ISCO with a smaller radius than a Schwarzschild BH by a factor of ≥1/6, depending on a for a prograde orbit: The ISCO radius approximately halves when the BH spin increases from $a = 0$ to 0.75. On the other hand, the orbital separation at the onset of mass shedding depends only weakly on the BH spin in the Boyer-Lindquist coordinates [11–13]. This decrease of the ISCO radius enhances the possibility for the disrupted material to escape capture by the BH and to form a more massive remnant disk than in the nonspinning BH case. The retrograde BH spin plays an opposite role; the ISCO radius of the Kerr BH increases by a factor of 1–1.5 for a retrograde orbit, and hence, the tidal effect is less important in the merger process.

Before closing this subsection, we estimate the degree of (undesired) orbital eccentricity in our simulations to assess the circularity of the orbital motion. For this purpose, we compute the evolution of the gauge-invariant orbital angular velocity $\Omega(t)$, which is defined from the $(l, m) = (2, 2)$ mode of Ψ_4 by

$$\Omega(t) = \frac{1}{2} \frac{|\Psi_4(l = m = 2)|}{|\int \Psi_4(l = m = 2)\mathrm{d}t|}. \tag{7.1}$$

The evolution of the orbital angular velocity in our simulation agrees with that derived from the Taylor-T4 formula in the inspiral phase within a small modulation, typically $\Delta\Omega/\Omega \lesssim 5\,\%$, which is equivalent to the orbital eccentricity of $\lesssim 3\,\%$. This amount of orbital eccentricity is as small as that observed in the nonspinning BH case with a low mass ratio $Q = 2$.

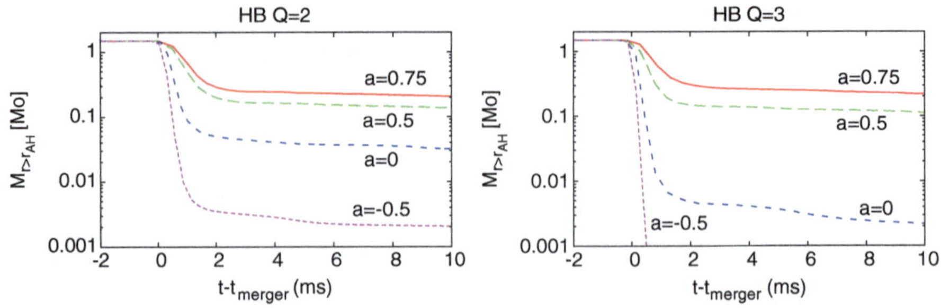

Fig. 7.5 Evolution of the rest mass of the material located outside an AH of the BH, $M_{r>r_{AH}}$. The *left* and *right* panels show the results for $Q = 2$ and 3, respectively. In both plots, $M_{NS} = 1.35M_\odot$ and HB EOS are adopted

7.3 Global Properties of the Disk

The mass of the remnant disk reflects the significance of NS tidal disruption in a clear way because the disk formation is a result of tidal disruption. A massive disk is formed if tidal disruption of the NS occurs far outside the ISCO. If the mass shedding starts in the vicinity of or inside the ISCO, only a small portion of the mass is left outside the AH. The material is not left outside the AH when the mass shedding does not occur before the BH swallows the NS, and the merger of a BH–NS binary may be indistinguishable from that of a BH–BH binary except for very small tidal corrections to the inspiral. Thus, the mass of a remnant disk is a reliable indicator of the degree of tidal disruption.

Figure 7.5 plots the time evolution of the rest mass located outside the AH, $M_{r>r_{AH}}$, for $Q = 2$ and 3 with different nondimensional BH spin parameters $a = 0.75, 0.5, 0$, and -0.5. In both plots, $M_{NS} = 1.35M_\odot$ and HB EOS are adopted. The dependence of $M_{r>r_{AH}}$ on a for HB EOS found here is similar to those for other EOSs. We set the time origin to be an approximate merger time t_{merger}. These plots indicate that the mass of the material left outside the AH relaxes to a quasisteady value for $t - t_{merger} \gtrsim 3$–4 ms, and the relaxed value increases monotonically as the BH spin increases from retrograde to prograde. This is consistent with the decrease of the BH ISCO radius with the increase of its spin, as described in Sect. 7.2. In particular, the remnant disk mass at $\approx$10 ms after the merger is $\gtrsim 0.1M_\odot$ for all the EOSs with $(Q, a) = (2, \geq 0.5)$ and $(\leq 4, 0.75)$, as shown in Table 7.5, and $\gtrsim 0.05M_\odot$ for $(Q, a) = (3, 0.5)$, irrespective of the EOS. The formation of such a massive disk may be encouraging for the BH–NS binary merger hypothesis of a short-hard GRB. For the $a = -0.5$ cases, by contrast, massive accretion disks of $\gtrsim 0.01M_\odot$ are not expected to be formed as merger remnants even for $Q = 2$ unless the EOS is extremely stiff (the NS radius is $\approx$15 km). This fact indicates that the retrograde BH spin is unfavorable for producing a central engine of a short-hard GRB.

The prograde BH spin enhances the disk formation dramatically for a BH–NS binary with *a high mass ratio*, for which the disk mass is very low when the BH is

Table 7.5 Several key quantities for the merger remnants for $M_{NS} = 1.35M_{\odot}$

Model	$M_{r>r_{AH}}[M_{\odot}]$	$C_e/4\pi M_0$	$M_{BH,c}/M_0$	$M_{irr,f}/M_0$	C_p/C_e	a_f	a_{f2}	a_{f1}
2H-Q2M135a75	0.32	0.913	0.915	0.789	0.807	0.87	0.87	0.95
1.5H-Q2M135a75	0.29	0.918	0.920	0.785	0.794	0.89	0.89	0.95
H-Q2M135a75	0.24	0.927	0.929	0.783	0.780	0.91	0.90	0.94
HB-Q2M135a75	0.21	0.933	0.934	0.783	0.772	0.91	0.91	0.94
B-Q2M135a75	0.18	0.937*	0.938	0.790*	0.778*	0.91*	0.91*	0.93
2H-Q2M135a5	0.27	0.925	0.926	0.825	0.843	0.81	0.81	0.84
1.5H-Q2M135a5	0.23	0.935	0.936	0.831	0.840	0.82	0.81	0.84
H-Q2M135a5	0.17	0.945	0.946	0.837	0.836	0.82	0.82	0.84
HB-Q2M135a5	0.14	0.951	0.952	0.840	0.832	0.83	0.83	0.84
B-Q2M135a5	0.095	0.959	0.960	0.846	0.830	0.83	0.83	0.84
2H-Q2M135a-5	0.13	0.961	0.962	0.931	0.954	0.48	0.48	0.50
H-Q2M135a-5	0.010	0.985	0.986	0.950	0.948	0.51	0.51	0.52
HB-Q2M135a-5	0.0021	0.985	0.986	0.952	0.950	0.50	0.50	0.51
B-Q2M135a-5	2×10^{-4}	0.983	0.984	0.952	0.952	0.49	0.49	0.50
2H-Q3M135a75	0.35	0.927	0.927	0.807	0.815	0.86	0.86	0.90
1.5H-Q3M135a75	0.30	0.931	0.934	0.811	0.815	0.86	0.86	0.90
H-Q3M135a75	0.24	0.939	0.943	0.820	0.818	0.85	0.85	0.91
HB-Q3M135a75	0.22	0.941	0.943	0.812	0.805	0.87	0.87	0.90
B-Q3M135a75	0.15	0.949	0.951	0.824	0.812	0.86	0.86	0.89
2H-Q3M135a5	0.28	0.939	0.940	0.858	0.874	0.74	0.74	0.77
1.5H-Q3M135a5	0.23	0.946	0.948	0.862	0.871	0.75	0.75	0.78
H-Q3M135a5	0.16	0.955	0.957	0.867	0.866	0.76	0.76	0.78
HB-Q3M135a5	0.11	0.961	0.963	0.871	0.864	0.77	0.77	0.78
B-Q3M135a5	0.050	0.969	0.971	0.877	0.862	0.77	0.77	0.79
HB-Q3M135a-5	$<10^{-4}$	0.986	0.987	0.973	0.980	0.32	0.32	0.33
2H-Q4M135a75	0.36	0.937	0.938	0.825	0.828	0.84	0.84	0.87
H-Q4M135a75	0.23	0.948	0.951	0.831	0.823	0.84	0.84	0.88
HB-Q4M135a75	0.18	0.953	0.956	0.833	0.821	0.85	0.85	0.88
B-Q4M135a75	0.11	0.960	0.963	0.837	0.817	0.85	0.85	0.88
2H-Q4M135a5	0.28	0.950	0.951	0.879	0.891	0.70	0.70	0.72
H-Q4M135a5	0.085	0.970	0.973	0.890	0.880	0.73	0.73	0.74
HB-Q4M135a5	0.024	0.976	0.979	0.894	0.878	0.74	0.74	0.75
B-Q4M135a5	0.0034	0.978	0.980	0.896	0.878	0.74	0.74	0.75
2H-Q5M135a75	0.36	0.946	0.947	0.838	0.835	0.82	0.82	0.85
H-Q5M135a75	0.17	0.960	0.963	0.844	0.827	0.84	0.84	0.86
HB-Q5M135a75	0.095	0.966	0.970	0.848	0.824	0.84	0.84	0.86
B-Q5M135a75	0.031	0.972	0.975	0.851	0.821	0.85	0.85	0.87

All the quantities are estimated at $\approx$10 ms after the approximate merger time $t = t_{merger}$. $M_{r>r_{AH}}$ is the rest mass of the disk surrounding the BH; because the accretion is still ongoing due to the hydrodynamic angular momentum transport process, the listed values only give approximate masses of the long-lived accretion disks, which survive for a time longer than the dynamical time scale $\sim$ a few ms. C_e and C_p are the circumferential radii of the AH along the equatorial and meridional planes, respectively. $C_e/4\pi$ and $M_{BH,c}$ denote approximate masses of the remnant BH. $M_{irr,f}$ is the irreducible mass of the remnant BH, and a_f is the nondimensional spin parameter of the remnant BH estimated from C_p/C_e. a_{f2} and a_{f1} are also the nondimensional spin parameters, estimated from the quantities on the AH and approximate conservation laws, respectively. We note that the values associated with the remnant BH for model B-Q2M135a75 (with an asterisk) are evaluated at $\approx$5 ms after the onset of the merger because the BH area decreases by $\gtrsim$1 % at $\approx$10 ms after the onset of the merger and the error becomes large

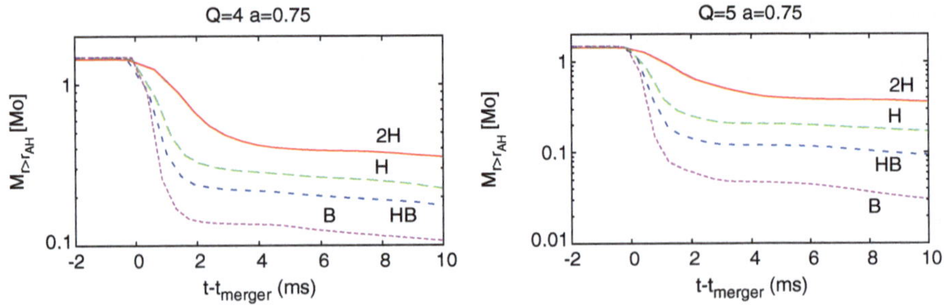

Fig. 7.6 The same as Fig. 7.5 but for different models $Q = 4$ (*left*) and 5 (*right*). In both plots, $M_{\rm NS} = 1.35M_\odot$ and $a = 0.75$ are adopted

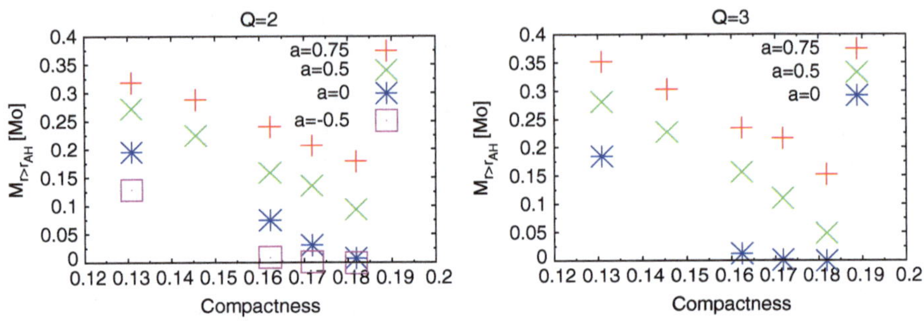

Fig. 7.7 The disk mass $M_{r>r_{\rm AH}}$ at ≈10 ms after the onset of the merger as a function of the NS compactness $\mathscr{C}$. The *left* and *right* panels show the results for $Q = 2$ (*left*) and $Q = 3$ (*right*)

nonspinning. We plot the time evolution of $M_{r>r_{\rm AH}}$ for $Q = 4$ and 5 with different EOSs in Fig. 7.6. In both plots, $M_{\rm NS} = 1.35M_\odot$ and $a = 0.75$ are adopted. Figure 7.6 clearly shows that a massive accretion disk is formed for $Q = 4$ and 5 if the BH has a prograde spin of $a = 0.75$. Namely, the formation of a massive accretion disk is universal for the merger of a BH–NS binary with a mass ratio of $Q \lesssim 5$ as far as $a \sim 0.75$ and $M_{\rm NS} = 1.35M_\odot$ (equivalently, $M_{\rm BH} \lesssim 6.75M_\odot$). Note that a heavy BH of $M_{\rm BH} \gtrsim 5M_\odot$ is predicted to be realistic as an astrophysical consequence of the stellar evolution with solar metallicity [14] (see, e.g., [15] for a population synthesis study) and hence as a possible progenitor of the short-hard GRB.

For more quantitative discussion, we plot the disk mass estimated at ≈10 ms after the merger for all the models with $Q = 2$ and for models with $(Q, a) = (3, \geq 0)$ as a function of the NS compactness, $\mathscr{C}$, in Fig. 7.7. Numerical values of $M_{r>r_{\rm AH}}$ are shown in Table 7.5, as well as other quantities associated with the merger remnants. For any fixed value of a, a negative correlation between $M_{r>r_{\rm AH}}$ and $\mathscr{C}$ is found to hold in Fig. 7.7. This correlation indicates that the NS with a larger compactness is less subject to tidal deformation and disruption than the NS with a smaller compactness for any fixed value of a. This correlation is expected from the nature of a tidal force as a finite-size effect, as found in the study of nonspinning BH–NS binaries. On

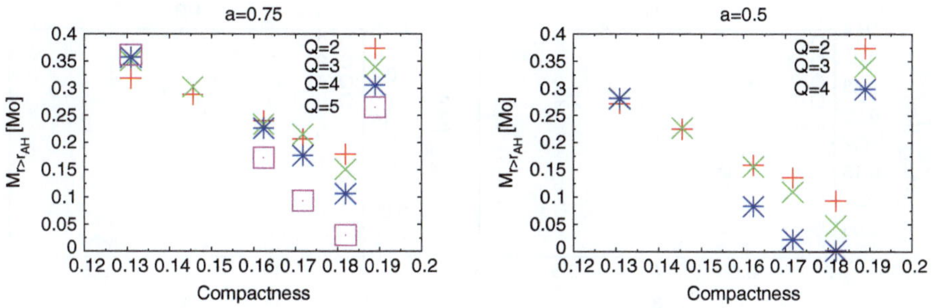

Fig. 7.8 The same as Fig. 7.7 but for $a = 0.75$ and $a = 0.5$ with $Q = 2 - 5$

the other hand, Fig. 7.7 again shows that the prograde BH spin increases the disk mass for any fixed value of $\mathscr{C}$. A remarkable fact is that the disk mass does not decrease steeply to a value of $\ll 0.1 M_\odot$ as the compactness increases for binaries with $(Q, a) = (\leq 3, \geq 0.5)$. We expect that the coalescence of a BH–NS binary with $(Q, a) = (\leq 3, \geq 0.5)$ may always produce a remnant disk of $\gtrsim 0.01 M_\odot$ within a plausible range of the NS compactness, $\mathscr{C} \lesssim 0.2$, although it is possible only if $\mathscr{C} \lesssim 0.18$ for $(Q, a) = (2, 0)$ and $\mathscr{C} \lesssim 0.16$ for $(Q, a) = (2, -0.5)$ or $(3, 0)$.

The dependence of the disk mass on the NS compactness is different for different values of the mass ratio. We plot in Fig. 7.8 the disk mass as a function of the NS compactness as in Fig. 7.7, but for $a = 0.75$ and 0.5. This figure shows that the disk mass depends more strongly on $\mathscr{C}$ when the mass ratio, Q, is larger. The disk mass is larger for smaller values of Q when the EOS is soft and $\mathscr{C} \gtrsim 0.16$, except for HB-Q2M135a75 and HB-Q3M135a75, for which the disk masses depend only weakly on Q. This dependence on Q is expected from the comparison between the mass-shedding radius, $r_{\rm shed}$, and the ISCO radius, $r_{\rm ISCO}$,

$$\frac{r_{\rm shed}}{r_{\rm ISCO}} \propto \mathscr{C}^{-1} Q^{-2/3}, \tag{7.2}$$

where we assume Newtonian gravity for simplicity. This relation states that a larger amount of mass can escape the capture by the BH and can form an accretion disk when Q is small because the mass shedding sets in at relatively more distant orbit. However, the disk mass may be larger for larger values of Q when the EOS is stiff as $\mathscr{C} \lesssim 0.15$ for $a \geq 0.5$ and $2 \lesssim Q \lesssim 5$. This should be ascribed to the redistribution process of the specific angular momentum of the NS to the disrupted material and to subsequent behavior of the material (such as collision of the fluid elements in spiral arms). This feature suggests that a binary with a larger value of Q, say $Q \gtrsim 6$, possibly form a massive remnant disk of $\gtrsim 0.1 M_\odot$ and could be a progenitor of a short-hard GRB if the EOS is stiff and the BH has a large spin $\gtrsim 0.5$.

To clarify the dependence of the disk mass on the BH spin, we plot the disk mass as a function of a in Fig. 7.9. The EOS (and, equivalently, $\mathscr{C}$) is the same for each plot. Again, we find a monotonic and steep increase of the disk mass as the increase

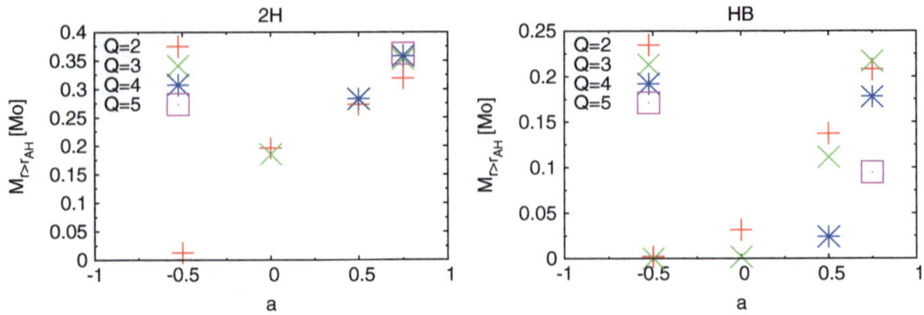

Fig. 7.9 The disk mass, $M_{r>r_{AH}}$, at $\approx$10 ms after the onset of the merger as a function of the nondimensional spin parameter of the BH a for models with $M_{NS} = 1.35M_\odot$. The *left* and *right* panels show the results for models with the 2H and HB EOSs, respectively

of a for the fixed EOS and mass ratio. The enhancement of the disk mass by a prograde spin is more dramatic for the compact NS (for the soft EOS). For example, the difference in the disk mass between the cases of $a = 0.75$ and -0.5 is only by a factor of $\sim$3 when $Q = 2$ and 2H EOS is adopted. This low amplification is natural because tidal disruption of a large NS occurs at an orbit far enough from the ISCO for a substantial amount of the disrupted material to escape the capture by the BH irrespective of a and because at such a large orbital separation the spin-orbit coupling effect is relatively weak. On the other hand, a few-orders-of-magnitude amplification of the disk mass is seen when $Q = 2$ and HB EOS is adopted.

Finally, we comment on a possible unbound outflow. To estimate the rest mass of unbound material, we compute

$$M_{\rm ub} \equiv \int_{r>r_{\rm AH}} \rho_* H(-u_t - 1) {\rm d}^3x, \tag{7.3}$$

where $H(x)$ is a step function. Here, the material with $u_t < -1$ should be considered to have an unbound orbit. We find that $M_{\rm ub}$ can be larger than $0.01M_\odot$ at $\approx$10 ms after the merger for the stiff EOS like 2H and H, and $a \geq 0$. However, $M_{\rm ub}$ does not approach a constant value and rather continues to decrease. Therefore, it is unclear whether $M_{\rm ub}$ estimated at 10 ms after the merger can really become unbound or not, and we do not show the precise values of $M_{\rm ub}$. When the EOS is not stiff, $M_{\rm ub}$ is negligible within the accuracy of our simulations.

7.4 Structure of the Remnant Disk

The structure of the remnant disk and its time evolution process depend on the mass ratio of the binary. We plot the rest-mass density profile at $\approx$5 and 10 ms after the onset of the merger for binaries with $a = 0.75$, HB EOS, and different values of Q in Fig. 7.10. The left column of Fig. 7.10 is plotted for $\approx$5 ms after the onset of

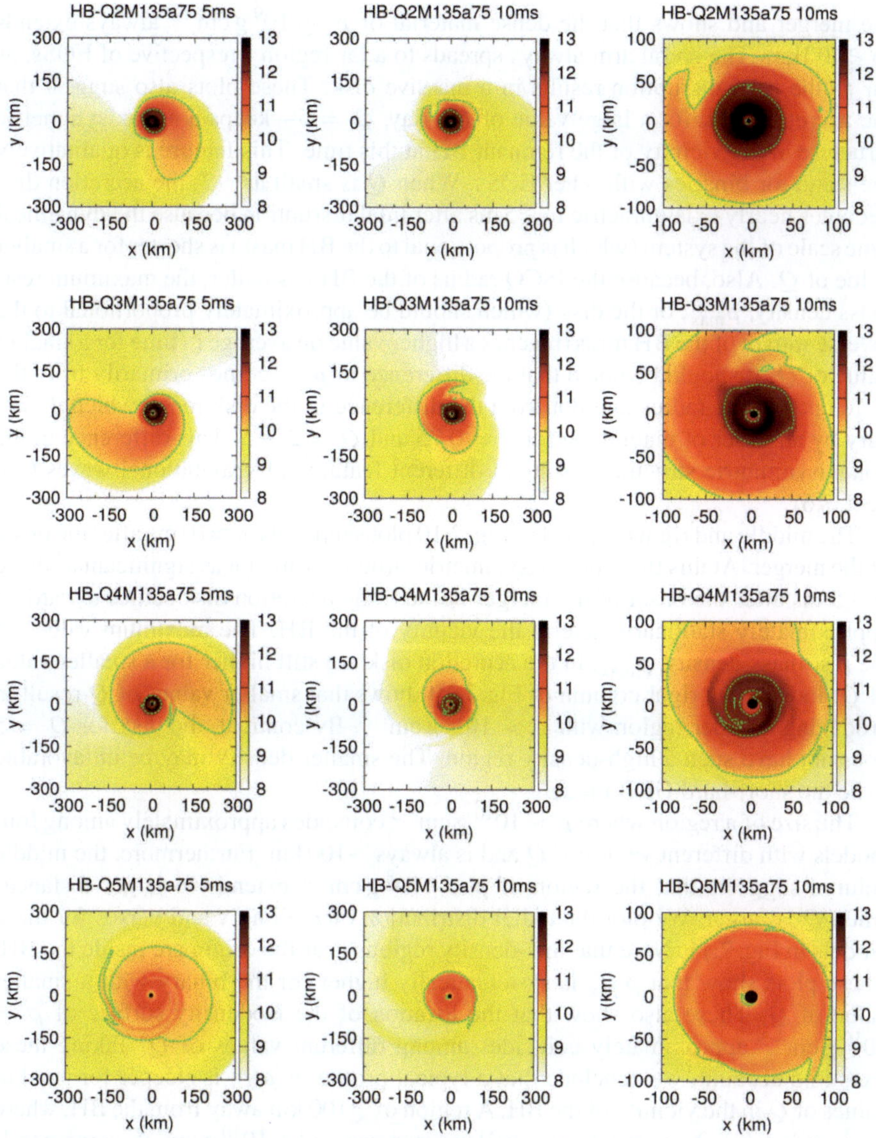

Fig. 7.10 The same as Fig. 7.2 with contour curves for $\rho = 10^{10}$ and $10^{12}\,\mathrm{g\,cm^{-3}}$ plotted. In all the plots, HB EOS and $a = 0.75$ are adopted. The first, second, third, and fourth rows are for $Q = 2, 3, 4$, and 5, respectively. The *left* column plots the snapshots at 5 ms after the onset of the merger. The *middle* column plots the snapshots at 10 ms after the onset of the merger, and the *right* column plots close-ups of the *middle* column

the merger and shows that the dense material of $\rho \gtrsim 10^9\,\mathrm{g\,cm^{-3}}$ always extends to $\gtrsim$400 km. The spiral arm always spreads to a far region irrespective of EOSs, as far as the tidal disruption results in a massive disk. These plots also suggest that the accretion disk for a large value of Q—say, $Q = 5$—keeps a nonaxisymmetric structure in the vicinity of the remnant BH at this time. This feature is qualitatively the same for binaries with other EOSs. When Q is small as $\sim$2, the accretion disk becomes nearly axisymmetric in $\approx$5 ms after tidal disruption because the dynamical time scale of the system (which is proportional to the BH mass) is shorter for a smaller value of Q. Also, because the ISCO radius of the BH is smaller, the maximum rest-mass density, ρ_{max}, of the disk (which should be approximately proportional to the inverse square of the BH mass) reaches a higher value on average in time for a smaller value of Q. It should be noted that the difference in ρ_{max} comes primarily from the difference in the radius and not from the difference in the disk mass, which do not vary by an order of magnitude for $a = 0.75$ and $Q = 2 - 5$. This difference in the nonaxisymmetric structure results in different features of gravitational waves (see Sect. 7.6).

The middle and right columns of Fig. 7.10 plot snapshots at $\approx$10 ms after the onset of the merger. At this time, nonaxisymmetric structures are not as significant as those at $\approx$5 ms after the onset of the merger because the accretion disk settles toward an approximately stationary state in the vicinity of the BH. The maximum values of the rest-mass density, ρ_{max}, in the accretion disk are still higher for a smaller value of Q. Indeed, the right column of Fig. 7.10 shows that smaller values of Q result in producing a wider region with $\rho > 10^{12}\,\mathrm{g\,cm^{-3}}$. By contrast, the disk for $Q = 5$ does not have such a high-density region. The smaller density may be unfavorable to be the short-hard GRB model.

The size of a region where $\rho > 10^{10}\,\mathrm{g\,cm^{-3}}$ coincides approximately among four models with different values of Q and is always $\sim$100 km. Furthermore, the middle column suggests that the region of $\rho > 10^8\,\mathrm{g\,cm^{-3}}$ extends to larger distances when Q is larger. We plot the radial distribution of ρ along x and y axes for these models in Fig. 7.11. Note that low-density regions near the origin are inside the BH. These plots show that ρ_{max} is systematically higher for the binary with a smaller value of Q. These also show that the location of the isodensity surface of $\rho = 10^{10}\,\mathrm{g\,cm^{-3}}$ approximately coincides among different values of Q. Taking these facts into account, we conclude that a typical profile of $\rho(r)$ is steeper for smaller values of Q in the vicinity of the BH. A region of $\gtrsim$100 km away from the BH, where the profile $\rho(r)$ shows relatively shallow decrease and $\lesssim 10^{10}\,\mathrm{g\,cm^{-3}}$, corresponds to the tail component, as is seen in the middle column of Fig. 7.10.

7.5 Properties of the Remnant BH

Table 7.5 shows that masses and nondimensional spin parameters of the remnant BHs depend weakly on the adopted EOSs. The mass of the remnant BH tends to become large as the EOS softens for fixed values of (Q, M_{NS}, a) for the case in which tidal

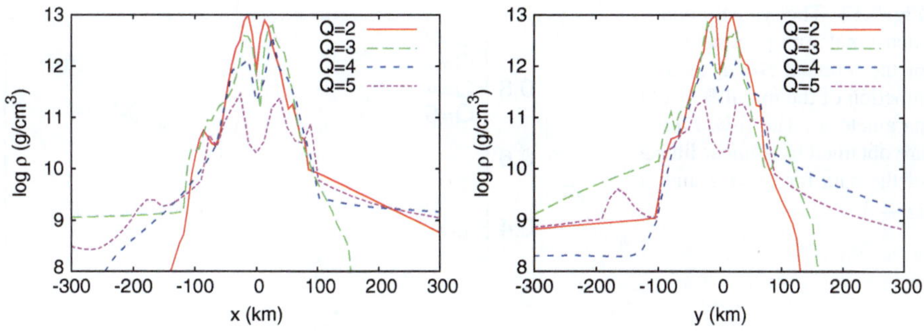

Fig. 7.11 Radial distribution of the rest-mass density at 10 ms after the onset of the merger for different values of Q. The *left* and *right* panels show the distribution along the x and y axes, respectively. In both plots, $(M_{\rm NS}, a) = (1.35M_\odot, 0.75)$ and HB EOS are adopted

disruption of the NS occurs. The reason for this is that the tidal disruption occurs near the ISCO, and then the BH swallows a large amount of the NS mass when the EOS is soft. Exceptionally, the mass of the BH becomes slightly larger for H and HB EOSs than for B EOS for binaries with $(Q, a) = (2, -0.5)$. The reason for this is that the remnant disk masses are small as $\lesssim 0.01M_\odot$ for these cases and the amount of the energy radiated by gravitational waves primarily determines the final state (for more compact NSs, the radiated energy is larger because a closer inspiral orbit is achieved). The spin angular momentum of the remnant BH $S_{\rm BH,f}$ shows similar behavior to that of the BH mass. The situation becomes complicated for a spin parameter of the remnant BH defined by $a_{\rm f} = S_{\rm BH,f}/M_{\rm BH,f}^2$; the competition between the mass and angular momentum losses from the system makes the dependence of the nondimensional spin parameter of the remnant BH on the EOS very weak. For comparison, $a_{\rm f1}$ and $a_{\rm f2}$ defined in Chap. 5 are also shown in Table 7.5. As is found in Chap. 6, $a_{\rm f}$ and $a_{\rm f2}$ agree with each other within the error of $\Delta a = 0.003$, and $a_{\rm f1}$ does not agree well with the other two estimates particularly when the massive remnant disk is formed and/or the mass of the BH is small: the maximum error is $\Delta a \approx 0.08$. A possible reason for this discrepancy is again ascribed to the underestimation of the angular momentum of the disk. Hereafter, we only refer to $a_{\rm f}$ as the nondimensional spin parameter of the remnant BH.

The nondimensional spin parameter of the remnant BH depends strongly on the initial spin parameter, a, and the mass ratio, Q. Approximate values of the nondimensional spin parameter of the remnant BH, $a_{\rm f}$, are shown in Fig. 7.12 as a function of the initial BH spin parameter, a. We also plot lines obtained by a linear fitting of data for $Q = 2$ and 3 of the following form,

$$a_{\rm f} = 0.32a + 0.66 \ (Q = 2), \tag{7.4}$$

$$a_{\rm f} = 0.43a + 0.54 \ (Q = 3). \tag{7.5}$$

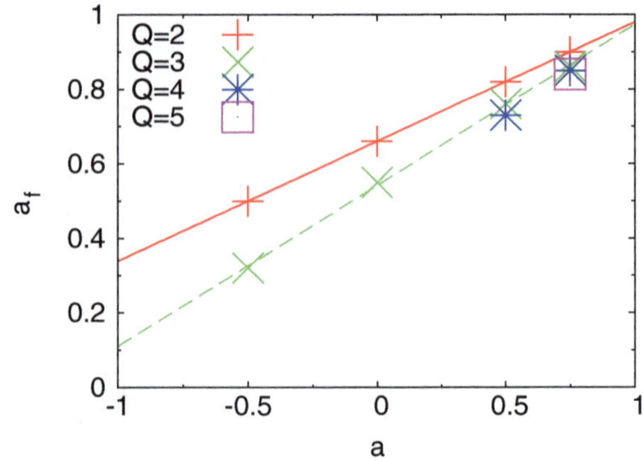

Fig. 7.12 The typical nondimensional spin parameters of the remnant BH a_{f} as a function of the initial BH spin parameter a. The *solid lines* are obtained by a linear fitting of the data for $Q = 2$ and $Q = 3$

The relation for $Q = 3$ agrees approximately with the results reported in [5, 16] within an error of $\lesssim 5\,\%$, and the agreement becomes better for a larger value of a. Figure 7.12 and these relations show that a_{f} is an approximately linear function of a. In a zeroth approximation, the slope and intercept of the linear relation denote the contribution from the initial BH spin angular momentum, S_{BH}, and the orbital angular momentum of the binary, J_0, respectively. The larger slope for a larger value of Q is explained by a larger contribution from the spin of the initial BH of mass $M_{\mathrm{BH}} = QM_{\mathrm{NS}}$ to the spin of the remnant BH of mass $M_{\mathrm{BH,f}} \sim (1+Q)M_{\mathrm{NS}}$. These predict the value of the slope to be $Q^2/(1+Q)^2$. However, the slope obtained by numerical simulations is smaller by $\sim$25–30 % than this predicted slope, because the amount of angular momenta redistributed to the remnant disk and extracted by gravitational waves become larger for a larger value of a. The fitting function also suggests that the merger of an extremely spinning BH of $a = 1$ and a NS with an irrotational velocity field results in a remnant BH with $a_{\mathrm{f}} \approx 0.98$ for BH–NS binaries with $Q = 2$ and 3 and hence never forms an overspinning BH, i.e., a BH with $a_{\mathrm{f}} > 1$. Furthermore, the results for $Q = 4$ shown in Table 7.5 also suggest $a_{\mathrm{f}} \approx 0.97$ for the merger of an extremely spinning BH and an irrotational NS. These results give a circumstantial support for cosmic censorship conjecture [17]. Whether $a_{\mathrm{f}} \lesssim 0.98(<1)$ is an universal consequence of a general BH–NS binary merger or not should be confirmed by simulations of higher mass-ratio binary mergers, in particular, with (nearly) extremal BH spin.

From these typical values of a_{f} and $M_{\mathrm{BH,f}}$, we can estimate typical frequencies of quasinormal modes (hereafter QNM) f_{QNM} of the remnant BH by a fitting formula [18]

$$f_{\mathrm{QNM}} M_{\mathrm{BH,f}} \approx \frac{1}{2\pi}[1.5251 - 1.1568(1-a_{\mathrm{f}})^{0.1292}]. \tag{7.6}$$

We plot these values in Fig. 7.13. They are in good agreement with those of the ringdown waveforms, if the QNM is excited after the merger.

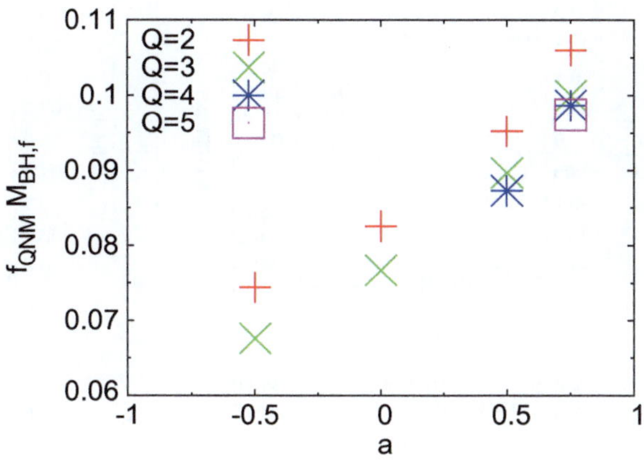

Fig. 7.13 The typical QNM frequency of the remnant BH normalized by its mass $f_{\rm QNM} M_{\rm BH,f}$

7.6 Gravitational Waveforms

In this section, we show $(l, m) = (2, 2)$, plus-mode gravitational waveforms h_+ for selected models obtained in this study. We plot all the waveforms for an observer along the z axis as a function of the approximate retarded time (the same as (6.2))

$$t_{\rm ret} = t - D - 2M_0 \ln(D/M_0). \tag{7.7}$$

The amplitude of the waveforms is normalized as Dh_+/m_0 or we show physical amplitude observed at a hypothetical distance $D = 100\,\text{Mpc}$ along the z axis. Gravitational waveforms calculated in the Taylor-T4 formula are plotted together in the figures to validate the waveforms obtained in our numerical simulations during the inspiral phase. Numerical waveforms during 2–3 initial cycles deviate from ones obtained from the Taylor-T4 formula in all the cases due to the lack of an approaching velocity in the initial data. This deficit is ascribed to insufficient modeling of the quasiequilibrium state and improvement in the future is important to obtain more accurate gravitational-wave templates [19, 20]. Our waveforms also deviate from the Taylor-T4 waveforms in the late inspiral phase due to a physical reason, which we describe below. Comparisons between waveforms obtained from simulations with different grid resolutions are shown in the Appendix of [9].

Figure 7.14 shows the gravitational waveforms for binaries with HB EOS, $(Q, M_{\rm NS}) = (3, 1.35M_\odot)$ but with different BH spin parameters, $a = 0.75$, 0.5, 0, and -0.5. This figure shows that the time to the merger, to which we refer approximately as the time at which the maximum gravitational-wave amplitude is achieved, for $\Omega_0 m_0 = 0.030$ becomes longer by $\approx$10 ms as the increase of the BH spin within the range concerned here. This difference in the merger time owes primarily to the spin-orbit interaction described in Sect. 7.2, and this behavior is qualitatively the same for binaries with any EOS. The numerical and Taylor-T4 waveforms agree well with each other during an inspiral phase for all the cases.

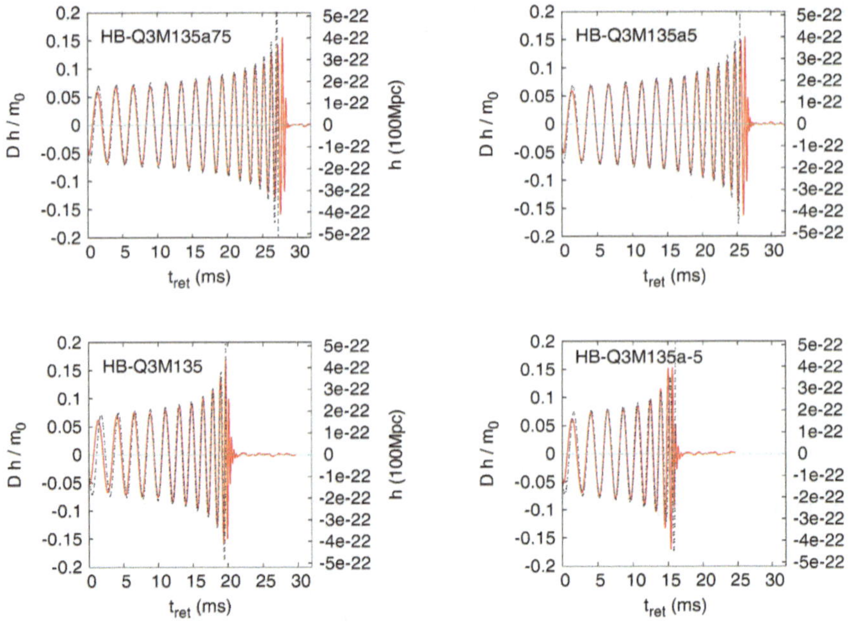

Fig. 7.14 $(l, m) = (2, 2)$, plus-mode gravitational waves for models HB-Q3M135a75, HB-Q3M135a5, HB-Q3M135, and HB-Q3M135a-5. All the waveforms are shown for an observer located along the z axis (the axis of rotation) and plotted as a function of a retarded time. The *left* axis denotes the amplitude normalized by the distance from the binary D and the total mass m_0. The *right* axis denotes the physical amplitude of gravitational waves observed at a hypothetical distance 100 Mpc. The *dotted curves* denote the waveform calculated by the Taylor-T4 formula

For the prograde BH spin cases, the Taylor-T4 formula does not track the evolution for $\sim$0.5 inspiral orbit just before the merger. The Taylor-T4 amplitude departs from that of numerical relativity and even diverges. Accordingly, the number of gravitational-wave cycles differs by as much as unity between the numerical and Taylor-T4 waveforms. The difference in the number of cycles is larger for higher mass-ratio binaries with prograde BH spins. We show the waveforms for binaries with $(Q, M_{\rm NS}, a) = (4, 1.35M_{\odot}, 0.75)$ and with $(Q, M_{\rm NS}, a) = (5, 1.35M_{\odot}, 0.75)$ for 2H, H, HB, and B EOSs in Figs. 7.15 and 7.16, respectively. The deviation is clear for H, HB, and B EOSs in both figures. This difference indicates that the phase evolution predicted by the Taylor-T4 formula is not sufficient to model the last inspiral phase of a coalescing binary with the high mass ratio of $Q \gtrsim 3$ and the prograde BH spin of $a \gtrsim 0.5$.

For the retrograde BH spin case (the bottom right panel of Fig. 7.14), the phase evolution deviates between the numerical and Taylor-T4 waveforms in the last orbit before the merger. This deviation may be partly ascribed to the small number of orbits in our simulation but appears to be primarily ascribed to a larger angular velocity, or equivalently a larger PN parameter, Ωm_0, at the last orbit for a retrograde BH

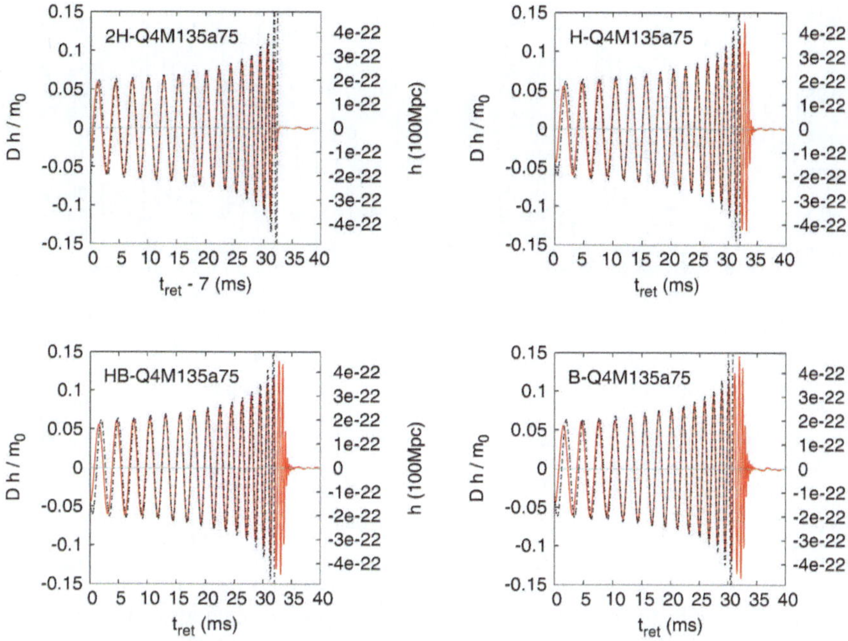

Fig. 7.15 The same as Fig. 7.14 but for models 2H-Q4M135a75, H-Q4M135a75, HB-Q4M135a75, and B-Q4M135a75

spin. Thus, the Taylor-T4 formula seems to be again insufficient for modeling the retrograde BH spin cases.

Figure 7.14 also shows that the gravitational waveform in the merger stage depends strongly on the BH spin. For a binary with $(Q, a) = (3, 0.75)$, gravitational waves show a sudden decrease in the amplitude at $t_{\rm ret} \approx 27$ ms, which is a clear signature of tidal disruption. Gravitational waves associated with the ringdown of a remnant BH are absent due to the phase cancellation by nearly axisymmetric accretion of the disrupted material. This feature is consistent with the formation of a massive remnant disk, which is described in Sect. 7.3, for the prograde BH spin. For binaries with $(Q, a) = (3, \leq 0)$, on the other hand, gravitational waves end up with ringdown waveforms associated with the remnant BHs because the tidal effect is very weak throughout the merger. In these circumstances, gravitational waves do not show strong signatures of tidal deformation and disruption of the NS.

Gravitational waves for a binary with $(Q, a) = (3, 0.5)$ show a qualitatively new feature (the top right panel of Fig. 7.14). In this case, a ringdown waveform of the remnant BH is seen in the final stage, although the NS is tidally disrupted and the disk mass is larger than $0.1 M_\odot$. Namely, both tidal disruption of the NS and excitation of a QNM of the remnant BH occur in a compatible manner. The same feature is also found for a binary with a high mass ratio and a prograde BH spin, i.e., $(Q, a) = (\geq 4, 0.75)$, shown in Figs. 7.15 and 7.16 except for 2H EOS, with

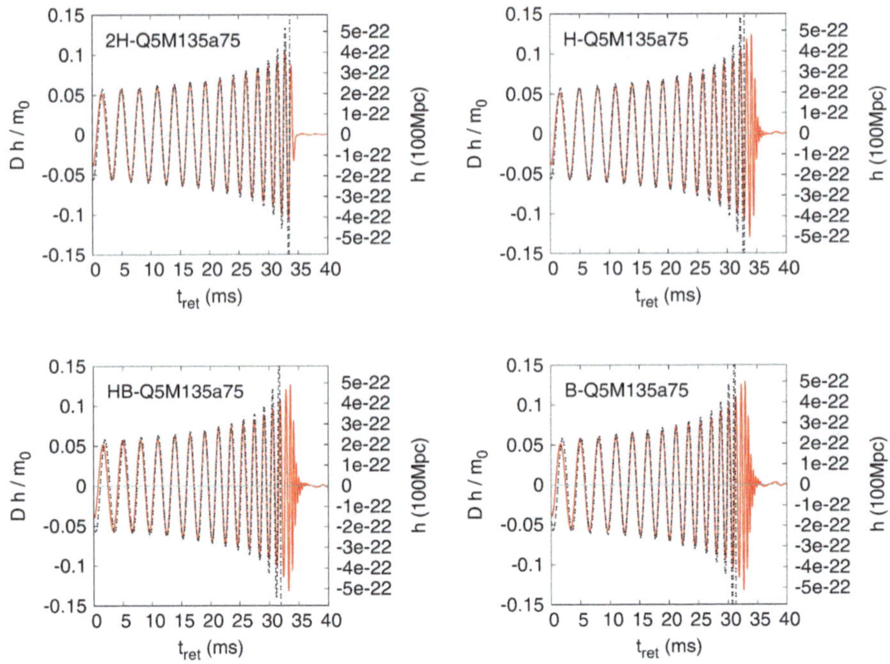

Fig. 7.16 The same as Fig. 7.14 but for models 2H-Q5M135a75, H-Q5M135a75, HB-Q5M135a75, and B-Q5M135a75

which the NS is disrupted at a fairly distant orbit. These waveforms are often seen for BH–NS binaries with a heavy BH (or a high mass ratio) with the prograde BH spin, which results in the NS tidal disruption, and is never seen for BH–NS binaries with $Q = 2$ or high mass-ratio binaries with nonspinning BHs.

The ratio of the areal radius of the remnant BH to the NS radius, $R_{\rm NS}$, is intimately related to the different excitation degree of the QNM between low and high mass-ratio binaries in the presence of NS tidal disruption [21, 22]. Schematic pictures of merger processes are depicted in Fig. 7.17. If tidal disruption does not occur, the NS is simply swallowed by the BH and excites a QNM, as shown in the middle panel of Fig. 7.17. If tidal disruption occurs in a binary with a low mass ratio, the disrupted material spreads around the BH to soon form a nearly axisymmetric disk. Approximately speaking, this occurs if the BH areal radius is smaller than the NS radius, as is shown in the left panel of Fig. 7.17. Thus, the NS tidal disruption has a strong effect to suppress the excitation of a QNM through the phase cancellation in the low mass-ratio binary. However, the situation is different in a high mass-ratio binary. Whereas the disrupted material forms an axisymmetric accretion disk around the BH in a sufficiently long time duration, the accretion just after the merger does not proceed in an axisymmetric way in high mass-ratio binaries, such as $Q = 4$, except for the extremely stiff EOS. This is because the BH radius for $Q = 4$ approximately

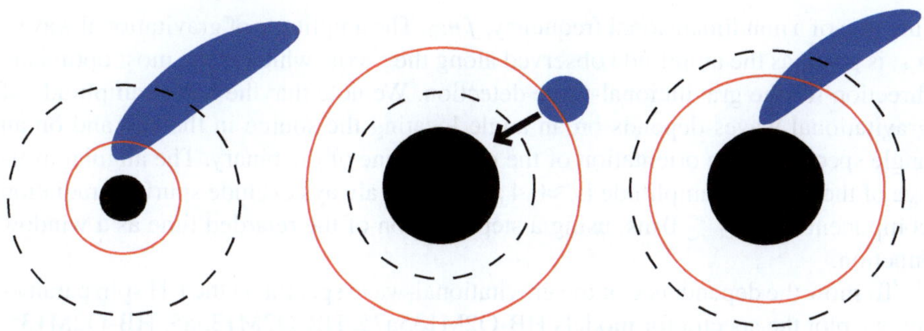

Fig. 7.17 Schematic pictures for three types of the merger process. The *solid filled circle* denotes the BH, the distorted ellipsoid denotes the NS, the *solid circle* is the location of the ISCO, and the *dashed circle* is the location of the radius at which the tidal disruption occurs. *Left* the NS is tidally disrupted, and the spatial extent of the disrupted material is larger than or as large as that of the BH. *Middle* the NS is not tidally disrupted. *Right* the NS is tidally disrupted, and the spatial extent of the disrupted material is smaller than that of the BH

doubles that for $Q = 2$, and hence, the disrupted material takes longer time to spread around the BH. In other words, the NS material accretes onto the BH coherently even after the tidal disruption, as is shown in the right panel of Fig. 7.17, because the BH radius is so large that the disrupted material cannot fully cover the BH surface before the BH swallows a large portion of the material. In the exceptional 2H EOS case, tidal disruption occurs sufficiently far outside the BH due to the large radius of the NS, and hence, the disrupted material is able to spread around the BH to form a nearly axisymmetric accretion disk before the prompt infall. Therefore, the QNM of a remnant BH is not excited for 2H EOS.

7.7 Gravitational-Wave Spectrum

Key features of gravitational waves are reflected in the Fourier spectrum. In this chapter, we define the Fourier spectrum as a sum of each Fourier component of two independent polarizations of the $(l, |m|) = (2, 2)$ mode as

$$\tilde{h}(f) = \sqrt{\frac{|\tilde{h}_+(f)|^2 + |\tilde{h}_\times(f)|^2}{2}}, \tag{7.8}$$

$$\tilde{h}_A(f) = \int e^{2\pi i f t} h_A(t) dt, \tag{7.9}$$

where A denotes two polarization modes, $+$ or $\times$. We show a nondimensional spectrum, $f\tilde{h}(f)$, observed at a hypothetical distance of 100 Mpc as a function of the gravitational-wave frequency, f (Hz), or a normalized amplitude, $f\tilde{h}(f)D/m_0$, as a function of a nondimensional frequency, fm_0. The amplitude of gravitational waves, h_A, is given as the amplitude observed along the z axis, which is the most optimistic direction for the gravitational-wave detection. We note that the actual amplitude of gravitational waves depends on an angle locating the source in the sky and on an angle specifying the orientation of the orbital plane of the binary. The angular average of the effective amplitude is $\approx 0.4 f\tilde{h}(f)$. We always exclude spurious radiation components for $t_{\rm ret} \lesssim 0\,{\rm ms}$, using a step function of the retarded time as a window function.

To show the dependence of the gravitational-wave spectra on the BH spin parameter, we plot the spectra for models HB-Q2M135a75, HB-Q2M135a5, HB-Q2M135, and HB-Q2M135a-5 in the left panel of Fig. 7.18 and for models HB-Q3M135a75, HB-Q3M135a5, HB-Q3M135, and HB-Q3M135a-5 in the right panel of Fig. 7.18. In the early inspiral phase of $f \lesssim 1\,{\rm kHz}$, where the point-particle approximation works well, the amplitude of the gravitational-wave spectrum for a given frequency increases monotonically as a increases. This is a feature expected from the PN calculation and is explained by the spin-orbit interaction as follows: The power spectrum of gravitational radiation is written as

$$\frac{{\rm d}E}{{\rm d}f} \propto [f\tilde{h}(f)]^2. \tag{7.10}$$

On the other hand, retaining only 1.5PN, the lowest-order spin-orbit interaction terms, (4.10) and (4.14) of [10], derive the expression for this quantity as

$$\frac{{\rm d}E}{{\rm d}f} = \frac{Q}{3(1+Q)^2}\frac{X^{5/2}}{\pi f^2}\left[1 + aX^{3/2}\left\{\frac{5(4Q+3)}{3(1+Q)^2}\right\}\right]. \tag{7.11}$$

Thus, the effective amplitude, $f\tilde{h}(f)$, for a given frequency f increases monotonically as the BH spin parameter, a, increases in the inspiral phase.

Figure 7.19 plots the spectra for models 2H-Q2M135a75, H-Q2M135a75, HB-Q2M135a75, and B-Q2M135a75, for which only the EOS is different, and indicates that the amplitude in the early inspiral phase does not depend strongly on $\mathscr{C}$. This is because the finite-size effect of the NS does not play an important role in the early inspiral phase (but, see [24, 25]), as already found for nonspinning BH–NS binaries.

In the late inspiral phase of 1 kHz $\lesssim f \lesssim f_{\rm cut}$, where $f_{\rm cut}$ is a characteristic frequency at which the spectrum starts damping exponentially (see below), the amplitude is significantly larger than the Taylor-T4 formula for the cases in which the NS is not disrupted. This is because the binaries in the inspiral and plunge after the NS enters the BH's ISCO emit gravitational waves in reality, whereas the Taylor-T4 formula does not take into account the motion inside the ISCO. In contrast to the spectrum calculated by the Taylor-T4 formula, which decreases steeply after the last inspiral phase, the amplitude obtained from the simulation depends only weakly on the gravitational-wave frequency in that phase, as far as the tidal disruption does not occur.

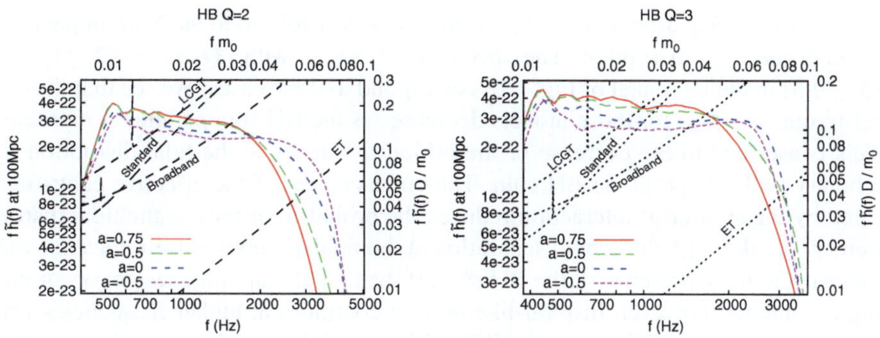

Fig. 7.18 Gravitational-wave spectra for BH–NS binaries with HB EOS, $M_{\rm NS} = 1.35M_\odot$ and $a = 0.75, 0.5, 0$, and -0.5. The *left* and *right* panels show the spectra for $Q = 2$ and 3, respectively. The *upper* axis denotes the normalized frequency, fm_0, and the *right* axis denotes the normalized amplitude, $f\tilde{h}(f)D/m_0$. The *bottom* axis denotes the frequency, f, in Hz, and the left axis denotes the nondimensional amplitude of gravitational waves, $f\tilde{h}(f)$, observed at a hypothetical distance 100 Mpc from the binary along the z axis. The *dashed curves* are planned noise curves of the LCGT ("LCGT"), the Advanced LIGO optimized for $1.4M_\odot$ NS-NS detection ("Standard"), the Advanced LIGO optimized for the burst detection ("Broadband"), and the Einstein Telescope ("ET") [23]

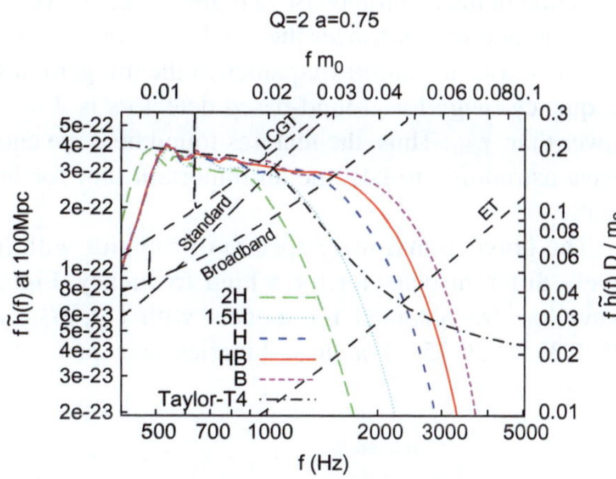

Fig. 7.19 The same as Fig. 7.18 but for $(Q, M_{\rm NS}, a) = (2, 1.35M_\odot, 0.75)$ with 2H, H, HB, and B EOSs. The spectrum derived by the Taylor-T4 formula is also included

The most fruitful information of the NS comes from the gravitational-wave spectrum in the merger phase through the "cutoff frequency," $f_{\rm cut}$, which depends on the BH spin as well as the NS compactness as described in Chap. 6 (see also [26]). If the NS tidal disruption occurs, the spectrum damps at $f = f_{\rm tidal} \sim 2$–4 kHz, which denotes the frequency at the tidal disruption and depends sensitively on physical parameters of the binary. In that case, gravitational waves for a higher frequency, $f \gtrsim f_{\rm cut} \approx f_{\rm tidal}$, are not emitted by the binary in the inspiral motion but only weakly by disrupted material. Because the disrupted material gradually spreads around the BH to form a nearly axisymmetric disk, the emission of gravitational waves is suppressed at the high frequency. Thus, the spectrum shows a relatively moderate damping around $f \approx f_{\rm cut}$, which is closely related to the NS compactness through the tidal disruption. The spectra for binaries with $(Q, a) = (2, \geq 0)$ and $(3, \geq 0.5)$ in the left panel of Fig. 7.18 correspond to these cases. We see that the cut-off frequency, $f_{\rm cut}$, for these models decreases as the BH spin parameter increases. This is ascribed to the decrease of the orbital frequency at the tidal disruption for a binary with the prograde BH spin. The enhancement of the effective centrifugal force by the spin-orbit interaction reduces the orbital frequency at the tidal disruption, $f_{\rm tidal}$, although the orbital separation at the tidal disruption itself does not vary much even in the presence of the BH spin. If the tidal disruption does not occur during the merger, however, inspiral-like motion continues at higher frequencies near and even inside the ISCO until the BH swallows the NS. In this case, the spectrum amplitude depends only weakly on f in the frequency range $f \lesssim f_{\rm cut}$ and damps for $f \gtrsim f_{\rm cut}$, which is closely related to the QNM frequency of the remnant BH, $f_{\rm QNM}$. The spectra for $(Q, a) = (2, -0.5)$ and $(3, \leq 0)$ in Fig. 7.18 show this feature. Note that the amplitude for model HB-Q3M135a-5 is smaller than for model HB-Q3M135 for the frequency range shown in Fig. 7.18 because tidal disruption does not play an important role, and (7.11) applies throughout the merger in both cases.

It is noteworthy that a prograde BH spin is favorable for the gravitational-wave detection in the inspiral phase and the estimation of $f_{\rm cut}$ in the merger phase because the prograde spin enhances the amplitude for a given frequency in the inspiral phase and decreases the cutoff frequency in the merger phase. Note that the most sensitive frequency range for ground-based detectors is $f \sim 10\,{\rm Hz}$–$1\,{\rm kHz}$, which is usually lower than $f_{\rm cut}$. Thus, the features found here are encouraging for the gravitational-wave astronomy to become an important tool for investigating the NS radius and EOS.

The gravitational-wave spectra of binaries with high mass ratios show qualitatively different behavior for a high frequency. Figure 7.20 plots the gravitational-wave spectra obtained for models with $(Q, M_{\rm NS}, a) = (4, 1.35M_\odot, 0.75)$ and $(5, 1.35M_\odot, 0.75)$. For these binaries (except for the model with 2H EOS), both

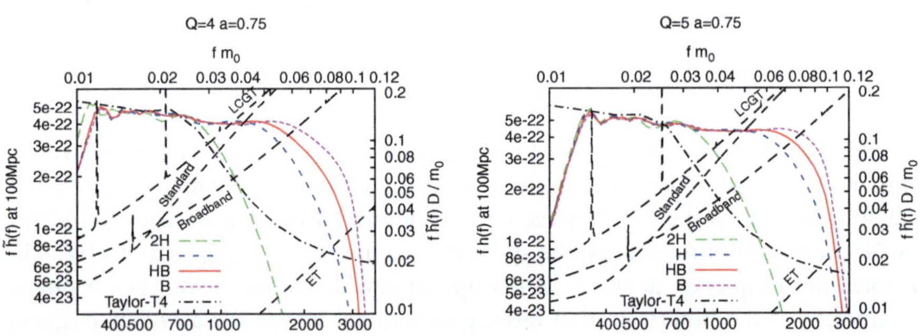

Fig. 7.20 The same as Fig. 7.18 but with the left panel for $(Q, M_{\rm NS}, a) = (4, 1.35M_\odot, 0.75)$ and the right panel for $(5, 1.35M_\odot, 0.75)$ with 2H, H, HB, and B EOSs. The spectrum derived by the Taylor-T4 formula is also included

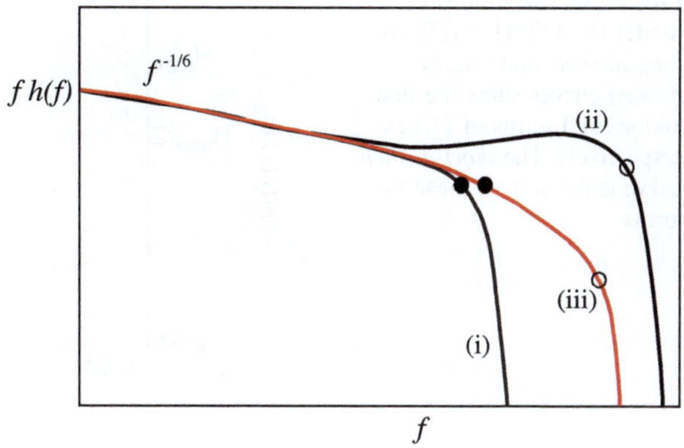

Fig. 7.21 A schematic figure of three types of gravitational-wave spectra. Spectrum *(i)* is for the case in which tidal disruption occurs far outside the ISCO, and spectrum *(ii)* is for the case in which tidal disruption does not occur. Spectrum *(iii)* is for the case in which tidal disruption occurs and the QNM is also excited. The *filled* and *open circles* denote $f_{\rm tidal}$ and $f_{\rm QNM}$, respectively

the NS tidal disruption and the excitation of a QNM of the remnant BH occur as is described in Sect. 7.6. Hence, the gravitational spectrum has two characteristic frequencies, i.e., $f_{\rm tidal}$ and $f_{\rm QNM}$, simultaneously. The spectra plotted in Fig. 7.20 indeed show such features. After the NS is tidally disrupted, the amplitude of the gravitational-wave spectrum shows a slow damp for $f \gtrsim f_{\rm tidal} \approx 2\,{\rm kHz}$. Then, the spectrum damps steeply above the frequency of the QNM, $f \gtrsim f_{\rm QNM} \approx 3\,{\rm kHz}$. A schematic figure of different spectra is depicted in Fig. 7.21, and the spectrum described in this paragraph corresponds to spectrum (iii) in this figure. This suggests that the cutoff frequency, $f_{\rm cut}$, of a high mass-ratio binary is not determined by a unique physical process like NS tidal disruption or a ringdown of a remnant BH, as far as both of them occur.

To estimate the cutoff frequency quantitatively, we fit the gravitational-wave spectra by a function with seven free parameters of the form (essentially the same as (6.6))

$$\frac{f\tilde{h}_{\rm fit}(f)D}{m_0} = \frac{f\tilde{h}_{\rm 3PN}(f)D}{m_0} e^{-(f/f_{\rm ins})^{\sigma_{\rm ins}}} + Ae^{-(f/f_{\rm dam})^{\sigma_{\rm dam}}}[1 - e^{-(f/f_{\rm ins2})^{\sigma_{\rm ins2}}}], \tag{7.12}$$

where $\tilde{h}_{\rm 3PN}(f)$ is the Fourier spectrum calculated by the Taylor-T4 formula. The first term in (7.12) models the inspiral spectrum, and the second term models the merger and ringdown spectra. We determine seven free parameters $f_{\rm ins}$, $f_{\rm ins2}$, $f_{\rm dam}$, $\sigma_{\rm ins}$, $\sigma_{\rm ins2}$, $\sigma_{\rm dam}$, and A by the condition that the following weighted norm is minimized (the same as (6.7)):

$$\sum_i \left\{[f_i\tilde{h}_i(f_i) - f_i\tilde{h}_{\rm fit}(f_i)]f_i^{1/3}\right\}^2. \tag{7.13}$$

Here, i denotes the data point for the spectrum. In Chap. 6 (see also [26]), we identify $f_{\rm dam}$ in (7.12) with $f_{\rm cut}$, which is most strongly correlated with the NS compactness

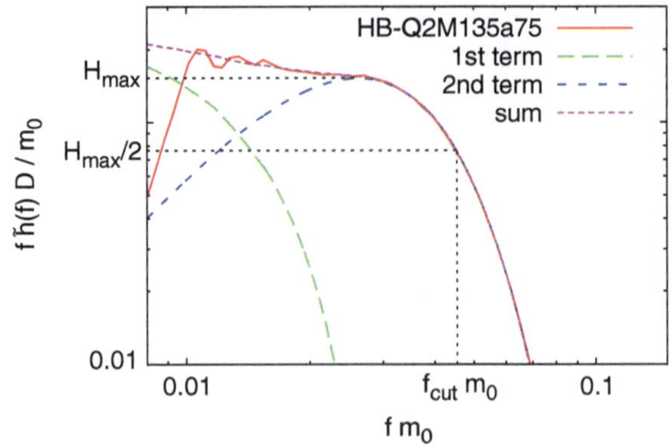

Fig. 7.22 The fitting for model HB-Q2M135a75. The *long-dashed* and *middle-dashed* curves show the first and second terms of (7.12), respectively. The *short-dashed* curve is the sum of these two terms

for nonspinning BH–NS binaries.[2] In this chapter, however, we obtain no strong correlation between $f_{\rm dam}$ (and the other parameters) and any parameter of physical importance, such as a or $\mathscr{C}$. The reason may be ascribed to the inadequacy of the functional form of (7.12), where the set of free parameters is degenerate to some extent. In particular, such a degeneracy is severe for a high mass-ratio binary due to two reasons. First, modeling an inspiral spectrum by the Taylor-T4 formula is inadequate for the late inspiral phase of a high mass-ratio binary due to the lack of information from the Taylor-T4 formula, as is described in Sect. 7.6. Second, there is no unique, physically motivated identification of $f_{\rm cut}$ when both the NS tidal disruption and the QNM excitation occur. (Fortunately, these degeneracies did not cause problems in the case of the nonspinning BH–NS binary with a low mass ratio in Chap. 6.) To overcome these problems with the fitting procedure, we redefine $f_{\rm cut}$ as the higher one of two frequencies at which the second term in (7.12) takes a half value of its maximum. An example of this fitting procedure is shown in Fig. 7.22. In this figure, $H_{\rm max}$ corresponds to the maximum value of the second term in (7.12). We find that this definition of $f_{\rm cut}$ works well to read off the NS compactness from the gravitational-wave spectrum.

Figure 7.23 shows $f_{\rm cut} m_0$ for spectra obtained for all binaries with $Q = 2$ as a function of the NS compactness, $\mathscr{C}$, in logarithmic scales. We also plot the typical QNM frequency of the remnant BH, $f_{\rm QNM}$, which depends primarily on a of the initial BH for a fixed value of Q. For each value of a, we find that $f_{\rm cut} m_0$ increases monotonically as $\mathscr{C}$ increases, and an approximate power law holds as

$$\ln(f_{\rm cut} m_0) \approx p(a) \ln \mathscr{C} + q(a) \quad (Q = 2), \tag{7.14}$$

where $p(a)(>0)$ and $q(a)$ depend only on a, for any value of a when $Q = 2$. This monotonic relation between $f_{\rm cut} m_0$ and $\mathscr{C}$ suggests us a possibility to extract the compactness, $\mathscr{C}$, of a NS from the gravitational-wave observation. It is noteworthy

[2] We refer to $f_{\rm dam}$ as $f_{\rm cut}$ throughout in Chap. 6. In this chapter, we distinguish $f_{\rm dam}$ from $f_{\rm cut}$ because the method for determining $f_{\rm dam}$ is different from that for $f_{\rm cut}$.

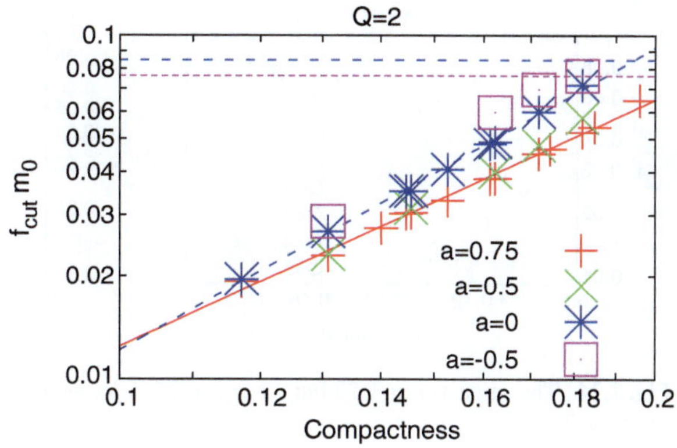

Fig. 7.23 The cutoff frequency times the total mass $f_{\rm cut} m_0$ as a function of the NS compactness $\mathscr{C}$ for $Q = 2$ binaries in logarithmic scales. The *solid* and *dashed lines* are obtained by linear fittings of data for $a = 0.75$ and $a = 0$, respectively. *Horizontal lines* denote the typical QNM frequencies of the remnant BHs. We note that $f_{\rm QNM} > f_{\rm cut}$ for $a = 0.75$ as long as $\mathscr{C} \leq 0.2$

that this relation includes only $\mathscr{C}$ but neither $M_{\rm NS}$ nor $R_{\rm NS}$ independently. It should also be noted that the simple relation found here is a consequence of our choice for a common value of the adiabatic index of the core EOS, $\Gamma_2 = 3$ (see Chap. 6). The increase of $f_{\rm cut} m_0$ with the increase of $\mathscr{C}$ indicates that a more compact NS is less subject to the BH tidal effect and disrupted at a closer orbit to the BH than a less compact NS is. The difference in $f_{\rm cut} m_0$ due to the difference in a becomes clearer for larger values of $\mathscr{C}$, and conversely, $f_{\rm cut} m_0$ depends only weakly on a if the compactness is as small as ≈ 0.12. The weak dependence on a for the small values of $\mathscr{C}$ is due to the fact that the effect of the BH spin at a distant orbit, at which the NS with a large radius is disrupted, is weak.

Figure 7.23 also shows that $p(a)$ is a decreasing function of a. More specifically we obtain the relations

$$\ln(f_{\rm cut} m_0) = (2.92 \pm 0.06) \ln \mathscr{C} + (2.32 \pm 0.12) \tag{7.15}$$

for $a = 0$[3] and

$$\ln(f_{\rm cut} m_0) = (2.39 \pm 0.06) \ln \mathscr{C} + (1.11 \pm 0.11) \tag{7.16}$$

for $a = 0.75$ by a linear fitting. The decreasing nature of $p(a)$ is explained by the fact that the spin-orbit repulsive force for the prograde BH spin, which reduces the orbital frequency at the NS tidal disruption, works efficiently for a close orbit and hence for the NS with a large value of $\mathscr{C}$. It is important that $p(a)$ is always larger than 1.5, which is expected from the analysis of the condition for the mass shedding,

$$\Omega m_0 \propto \frac{\mathscr{C}^{3/2}(1+Q)^{3/2}}{\sqrt{Q}}. \tag{7.17}$$

[3] The relation between $f_{\rm cut} m_0$ and $\mathscr{C}$ is different from the one obtained in Chap. 6 due to the different definition of $f_{\rm cut}$.

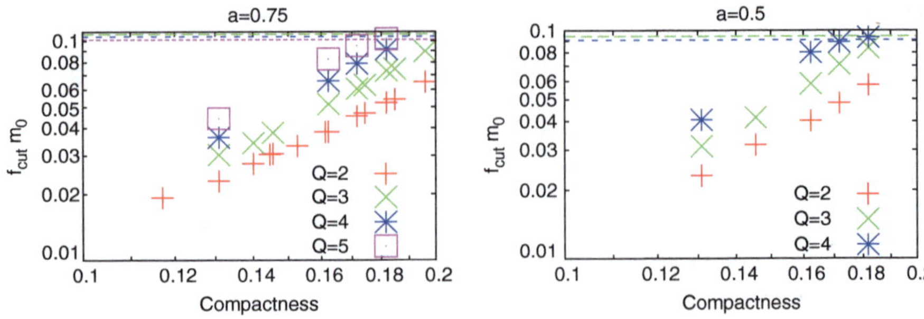

Fig. 7.24 The same as Fig. 7.23 but for $a = 0.75$ (*left*) and 0.5 (*right*). In both panels, $f_{\rm QNM} > f_{\rm cut}$ for $Q = 2$

The large value of $p(a)$ is favorable for determining the NS compactness from the gravitational-wave observation because the dependence of $f_{\rm cut}$ on $\mathscr{C}$ becomes stronger. Note that $f_{\rm cut}$ is always lower than the QNM frequency of the remnant BH for a realistic range of the compactness $\mathscr{C} \lesssim 0.2$ for $(Q, a) = (2, \gtrsim 0)$. If a is negative, on the other hand, $f_{\rm cut}$ for the binary with a compact NS of $\mathscr{C} \gtrsim 0.18$ may be determined by the QNM frequency, $f_{\rm QNM}$, and it will be difficult to determine the NS compactness from the cutoff frequency.

Figure 7.24 shows the $f_{\rm cut} m_0$–$\mathscr{C}$ relation of gravitational-wave spectra obtained for all binaries with $a = 0.75$ and $a = 0.5$. This figure, combined with Fig. 7.23, clearly indicates that the approximate power law of the form

$$\ln(f_{\rm cut} m_0) = p(Q, a) \ln \mathscr{C} + q(Q, a) \tag{7.18}$$

holds for binaries of $\mathscr{C} \lesssim 0.2$ with $Q = 5$ as far as $a \sim 0.75$ and with $Q \leq 4$ as far as $a \sim 0.5$. The striking feature is that the cutoff frequency is lower than the QNM frequency of the remnant BH, $f_{\rm QNM}$, for $(Q, a) = (\leq 4, 0.75)$ and for $(Q, a) = (\leq 3, 0.5)$ even if a QNM is excited. Accordingly, $f_{\rm cut}$ shows a strong correlation with $\mathscr{C}$. For $(Q, a) = (5, 0.75)$ and (4, 0.5), $f_{\rm cut}$ is lower than $f_{\rm QNM}$ as far as $\mathscr{C} \lesssim 0.18$ and 0.17, respectively, and, therefore, the strong correlation between $f_{\rm cut}$ and $\mathscr{C}$ is found within this range. Although $f_{\rm cut}$ for the binary with a high mass ratio should not be considered as $f_{\rm tidal}$ due to the QNM excitation, monotonic relations between $f_{\rm cut} m_0$ and $\mathscr{C}$ gives us an opportunity to explore the NS radius and EOS. It should be noted that gravitational waves from a higher mass-ratio binary are more subject to the gravitational-wave detection due to the larger amplitude in the inspiral phase and the lower cutoff frequency. We again note that a massive BH of $M_{\rm BH} \gtrsim 5 M_\odot$ is an astrophysically realistic consequence of the stellar evolution [14, 15]. Taking these facts into account, we conclude that gravitational waves from the BH–NS binary are a promising tool to investigate the NS radius and EOS in the next decade.

7.8 Energy and Angular Momentum Radiated by Gravitational Waves

Tables 7.6 and 7.7 list the total energy $\Delta E/M_0$ and angular momentum $\Delta J/J_0$ radiated by gravitational waves. We estimate systematic errors in the estimation of ΔE and ΔJ to be $\sim$10 %, which are ascribed mainly to the finite grid resolution and partly to the finite extraction radius for Ψ_4. Because ΔE and ΔJ depend on the choice of $\Omega_0 m_0$, we do not compare directly the results obtained for models with different values of $\Omega_0 m_0$ and accordingly models with different values of Q (see Tables 7.1 and 7.2).

Contributions from all the $l = 2$–4 modes of gravitational waves are taken into account. The $(l, |m|) = (2, 2)$ mode always contributes by $\gtrsim$85 % to ΔE and ΔJ. ΔE and ΔJ taken away by higher-mode gravitational waves are substantial for high mass-ratio binaries. For example, the (3, 3) mode contributes by $\sim$ 2, 5, 7.5, and 10 % for binaries with $Q = 2, 3, 4,$ and 5, respectively. The (4, 4) mode gravitational waves contribute by $1 \sim 2$ % for binaries with $Q = 3$–5. These values depend only weakly on a and the EOS, and contributions of modes with $l \neq m$ are negligible compared to those of $l = m$ modes.

Tables 7.6 and 7.7 show that $\Delta E/M_0$ and $\Delta J/J_0$ increase monotonically as the NS compactness, $\mathscr{C}$, increases for binaries with fixed values of (Q, a). This is the same result as that obtained for nonspinning BH–NS binaries and is explained by a longer inspiral phase for a softer EOS due to the later onset of mass shedding and the later tidal disruption. The ratio between these two values, $(\Delta J/J_0)/(\Delta E/M_0)$, decreases as $\mathscr{C}$ increases. This agrees again with the result for the nonspinning BH cases and is explained by a relation $\Delta J/\Delta E \approx m/\Omega$ for a given angular harmonic of m and by the fact that more radiation is emitted from the orbit of a larger value of Ω for a softer EOS. Note that these arguments are based on little dependence of gravitational-wave luminosity in the inspiral phase on $\mathscr{C}$ for a fixed value of a; tidal correction to the luminosity in the inspiral phase is not important.

Table 7.6 shows that $\Delta E/M_0$ does not depend strongly on a, while $\Delta J/J_0$ increases as a increases in many cases for a fixed value of $\mathscr{C}$. Remember that $\mathrm{d}E/\mathrm{d}f$ in the inspiral phase increases for a large value of a, as is given by (7.11). However, the orbital frequency, Ω, at the tidal disruption decreases for a large value of a due to the spin-orbit interaction. Because of these two competing effects, the binding energy at the tidal disruption depends only weakly on a, and hence, $\Delta E/M_0$ does not change very much among different values of a. The increase of $\Delta J/J_0$ for a large value of a is due to the large value of $\mathrm{d}E/\mathrm{d}f$ in the inspiral phase, during which Ω is relatively low, and to the approximate relation $\Delta J \approx m\Delta E/\Omega$, which enhances the contribution of low-frequency gravitational waves.

Finally, we comment on the linear momentum ΔP radiated by gravitational waves and an associated kick velocity $v_{\mathrm{kick}} \equiv \Delta P/M_0$ of the remnant BH. Because of the mass and spin asymmetries, the remnant BH achieves the kick velocity of $\sim$100–250 km s^{-1} when the effect of tidal disruption is weak, e.g., for models HB-Q3M135a-5 and B-Q3M135. Although our results for ΔP do not converge as well

Table 7.6 Total radiated energy ΔE and angular momentum ΔJ carried away by gravitational waves for $Q = 2$ and 3

Model	$\Delta E/M_0$ (%)	$\Delta J/J_0$ (%)	$(\Delta J/J_0)/(\Delta E/M_0)$
2H-Q2M135a75	0.58	16	27
1.5H-Q2M135a75	0.79	19	24
H-Q2M135a75	1.1	24	21
HB-Q2M135a75	1.4	26	19
B-Q2M135a75	1.7	29	17
2H-Q2M135a5	0.60	17	26
1.5H-Q2M135a5	0.79	19	24
H-Q2M135a5	1.2	24	20
HB-Q2M135a5	1.4	26	19
B-Q2M135a5	1.7	28	17
2H-Q2M135a-5	0.57	15	26
H-Q2M135a-5	1.1	19	16
HB-Q2M135a-5	1.4	19	14
B-Q2M135a-5	1.6	21	13
2H-Q2M12a75	0.40	12	30
H-Q2M12a75	0.79	19	24
HB-Q2M12a75	0.95	21	22
B-Q2M12a75	1.2	24	21
2H-Q2M145a75	0.73	19	25
H-Q2M145a75	1.5	27	19
HB-Q2M145a75	1.7	30	17
B-Q2M145a75	2.1	32	15
2H-Q3M135a75	0.72	20	28
1.5H-Q3M135a75	0.97	23	24
H-Q3M135a75	1.3	27	20
HB-Q3M135a75	1.6	30	19
B-Q3M135a75	2.0	34	17
2H-Q3M135a5	0.70	19	27
1.5H-Q3M135a5	0.94	22	23
H-Q3M135a5	1.4	26	19
HB-Q3M135a5	1.7	29	17
B-Q3M135a5	2.0	31	15
HB-Q3M135a-5	1.3	19	14
2H-Q3M145a75	0.88	22	25
H-Q3M145a75	1.7	31	18
HB-Q3M145a75	2.1	34	16
B-Q3M145a75	2.5	37	15

ΔE and ΔJ are normalized with respect to the initial ADM mass M_0 and angular momentum J_0, respectively. We also show the ratio between ΔJ and ΔE. (See Table 6.3 in Chap. 6 for models of nonspinning BH–NS binaries)

Table 7.7 The same as Table 7.6 but for $Q = 4$ and 5

Model	$\Delta E/M_0$ (%)	$\Delta J/J_0$ (%)	$(\Delta J/J_0)/(\Delta E/M_0)$
2H-Q4M135a75	0.81	23	28
H-Q4M135a75	1.5	31	21
HB-Q4M135a75	1.8	33	19
B-Q4M135a75	2.1	36	17
2H-Q4M135a5	0.72	19	27
H-Q4M135a5	1.5	27	19
HB-Q4M135a5	1.7	29	17
B-Q4M135a5	1.9	31	16
2H-Q5M135a75	0.83	24	29
H-Q5M135a75	1.6	33	20
HB-Q5M135a75	1.9	35	19
B-Q5M135a75	2.1	36	17

as those for ΔE and ΔJ due to the slow convergence of $(l, m) \neq (2, 2)$ mode gravitational waves, the values of $v_{\rm kick}$ are in reasonable agreements with the fitting formula derived using the results of simulations for the binary BH merger [27, 28]. By contrast, $v_{\rm kick}$ is suppressed to $\lesssim$100 km s^{-1} when tidal disruption occurs far outside the ISCO. The reason for this is that the tidal disruption suppresses significantly the gravitational radiation from the last inspiral and merger phases, during which the linear momentum is emitted most efficiently. This trend is consistent with the result found in [26].

References

1. B.D. Lackey, K. Kyutoku, M. Shibata, P.R. Brady, J.L. Friedman, Phys. Rev. D **85**, 044061 (2012)
2. M. Shibata, Prog. Theor. Phys. **96**, 917 (1996)
3. P. Wiggins, D. Lai, Astrophys. J. **532**, 530 (2000)
4. J.M. Bardeen, W.H. Press, S.A. Teukolsky, Astrophys. J. **178**, 347 (1972)
5. Z.B. Etienne, Y.T. Liu, S.L. Shapiro, T.W. Baumgarte, Phys. Rev. D **79**, 044024 (2009)
6. L. Blanchet, Living Rev. Relativ. **9**, 4 (2006)
7. K. Kyutoku, M. Shibata, K. Taniguchi, Phys. Rev. D **82**, 044049 (2010)
8. K. Kyutoku, M. Shibata, K. Taniguchi, Phys. Rev. D **84**, 049902(E) (2011)
9. K. Kyutoku, H. Okawa, M. Shibata, K. Taniguchi, Phys. Rev. D **84**, 064018 (2011)
10. L.E. Kidder, Phys. Rev. D **52**, 821 (1995)
11. L.G. Fishbone, Astrophys. J. **185**, 43 (1973)
12. J.A. Marck, Proc. R. Soc. Lond. **385**, 431 (1983)
13. M. Ishii, M. Shibata, Y. Mino, Phys. Rev. D **71**, 044017 (2005)
14. J.E. McClintock, R.A. Remillard, in *Compact Stellar X-ray Sources*, ed. by W.H.G. Lewin, M. van der Klis (Cambridge University Press, Cambridge, 2006)
15. K. Belczynski, T. Bulik, C.L. Fryer, A. Ruiter, F. Valsecchi, J.S. Vink, J.R. Hurley, Astrophys. J. **714**, 1217 (2010)
16. F. Foucart, M.D. Duez, L.E. Kidder, S.A. Teukolsky, Phys. Rev. D **83**, 024005 (2011)

17. R. Penrose, in *General Relativity: An Einstein Centenary Survey*, ed. by S.W. Hawking, W. Israel (Cambridge University Press, Cambridge, 1979)
18. E. Berti, V. Cardoso, A.O. Starinets, Class. Quantum Gravity **26**, 163001 (2009)
19. K. Uryū, F. Limousin, J.L. Friedman, E. Gourgoulhon, M. Shibata, Phys. Rev. Lett. **97**, 171101 (2006)
20. K. Uryū, F. Limousin, J.L. Friedman, E. Gourgoulhon, M. Shibata, Phys. Rev. D **80**, 124004 (2009)
21. M. Saijo, T. Nakamura, Phys. Rev. Lett. **85**, 2665 (2000)
22. M. Saijo, T. Nakamura, Phys. Rev. D **63**, 064004 (2001)
23. M. Punturo et al., Class. Quantum Gravity **27**, 194002 (2010)
24. L. Baiotti, T. Damour, B. Giacomazzo, A. Nagar, L. Rezzolla, Phys. Rev. Lett. **105**, 261101 (2010)
25. L. Baiotti, T. Damour, B. Giacomazzo, A. Nagar, L. Rezzolla, Phys. Rev. D **84**, 024017 (2011)
26. M. Shibata, K. Kyutoku, T. Yamamoto, K. Taniguchi, Phys. Rev. D **79**, 044030 (2009)
27. L. Rezzolla, Class. Quantum Gravity **26**, 094023 (2009)
28. Y. Zlochower, M. Campanelli, C.O. Lousto, Class. Quantum Gravity **28**, 114015 (2011)

Chapter 8
Summary

We performed numerical simulations of the BH–NS binary merger for a wide range of binary parameters using a fully general relativistic AMR code, SACRA, with nine piecewise polytropic EOSs. In this work, we employed the EOSs with two free parameters which determine the core EOSs. The crust EOS was fixed throughout the study, and the core EOS was varied for a wide range to perform systematic investigation. We investigated the dependence of the merger process, properties and structures of the remnant disk, properties of the remnant BH, gravitational waveforms, and their spectra on the NS EOS. In particular, we focused on the case in which the mass ratio is small and/or the BH has a prograde spin, and the tidal disruption of the NS by a companion BH plays an important role. We adopted a number of parameters for the mass ratio, NS mass, and BH spin, which is zero or (anti)aligned with the orbital angular momentum of a binary. By preparing the initial condition with a distant orbit and a small eccentricity, we always tracked $\gtrsim$5 quasicircular orbits in the inspiral phase and studied the merger phase in a realistic setting. We also evolved the merger remnant, i.e., the BH-disk system, until they settled to a quasistationary state. In the following, we summarize the conclusions of this thesis.

8.1 Summary of Our Results

First, the conclusion for the remnant disk and BH is as follows:

1. The disk mass depends strongly on the EOS, because the EOS determines the location at which the tidal disruption occurs through the compactness $\mathscr{C}$ of the NS. For a nonspinning BH–NS binary, the disk mass is correlated strongly with the NS compactness $\mathscr{C}$, and for $Q = 2$, it can be $\gtrsim 0.01 M_\odot$ for a wide range of the EOSs and the NS masses. However, the disk mass is tiny for $Q = 3$, unless the EOS is extremely stiff like 2H EOS or the NS mass is low. For the BH-NS binaries consisting of a nonspinning BH, the disk mass can be $\gtrsim 0.01 M_\odot$ for $Q = 3$, only for the case in which $\mathscr{C} \lesssim 0.16$.

K. Kyutoku, *The Black Hole–Neutron Star Binary Merger in Full General Relativity*, Springer Theses, DOI: 10.1007/978-4-431-54201-8_8, © Springer Japan 2013

2. It is shown that a prograde BH spin enhances the effect of NS tidal disruption by the spin-orbit interaction. The mass of the remnant disk increases as the BH spin increases, because the ISCO radius of the BH becomes small. A remarkable point is that the BH-NS binary with a high mass ratio of even $Q = 5$ can form a sufficiently massive disk of $\gtrsim 0.1 M_\odot$ for a wide range of the NS compactness if the BH has a prograde spin of $a = 0.75$. This amount of the disk mass for a high mass-ratio binary is hardly expected if the BH is nonspinning. This fact suggests that the formation of a BH-massive accretion disk system is a frequent outcome of the BH-NS binary merger with a prograde BH spin and, may be encouraging for the merger scenario of a short-hard GRB. By contrast, the disk mass becomes very small if the BH has a retrograde spin.
3. It is shown that some portion of the disrupted material can extend to $\gtrsim 400\,\mathrm{km}$ from the BH if the massive disk is formed. The maximum rest-mass density in the disk is larger for binaries with smaller values of Q, because the ISCO radius and the length scale of the system are smaller for such binaries. The extent of the disk could be larger for a larger value of Q. For such a remnant disk, the lifetime should be longer.
4. The spin parameter of the remnant BH depends primarily on the spin parameter of the initial BH, a, and the mass ratio, Q. By contrast, the spin parameter depends only weakly on the EOS for given masses of the BH and NS, unlike the disk mass. In particular, extrapolation of our results suggests that the merger of an extremely spinning BH and an irrotational NS does not form an overspinning BH.

Next, the conclusion for gravitational waves is as follows:

1. For the case in which the tidal disruption occurs before the orbit reaches the ISCO, the gravitational-wave amplitude decreases quickly at its onset, and the emission of ringdown gravitational waves associated with the quasinormal mode of the remnant BH is suppressed for a nonspinning BH–NS binary. Only in the nonspinning BH–NS binaries with low values of mass ratio, the tidal effects play an important role, and hence the remarkable dependence of the gravitational waveforms on the EOS is found only for such cases. With stiffer EOSs, the radius of the NS becomes larger and the tidal effect is more relevant than with softer EOSs.
2. The gravitational waveform also depends strongly on the BH spin. The number of gravitational-wave cycles becomes larger for a prograde BH spin than that for a zero BH spin in the inspiral phase, because an additional repulsive force due to the spin-orbit interaction reduces gravitational-wave luminosity and an approaching velocity of the binary. We found that the Taylor-T4 formula does not reproduce the phase evolution in the late inspiral phases accurately, especially when the mass ratio is large.
3. For nonspinning BH–NS binaries, the waveforms are classified simply into two categories. When tidal disruption of the NS occurs, the waveform is composed of an inspiral waveform and a prompt shutdown at the tidal disruption. When tidal disruption does not occur, the waveform is composed of inspiral and QNM

waveforms. However, we find that the NS tidal disruption and the excitation of a QNM can occur simultaneously for binaries with a high mass ratio and a prograde BH spin. This is because the disrupted material cannot become axisymmetric before the prompt infall due to a larger BH areal radius for a larger value of Q. As a result, the material accretes onto the remnant BH coherently, and therefore the QNM of the remnant BH is excited, except for the case in which the extremely stiff EOS is adopted.

4. The Fourier spectrum of gravitational waves is characterized by a cutoff frequency, $f_{\rm cut}$, above which the spectrum amplitude exponentially damps. We find that the cutoff frequency $f_{\rm cut}$ depends strongly on the mass ratio and the compactness $\mathscr{C}$ of the NS. For a given small mass ratio such as $Q = 2$, the value of $f_{\rm cut}$ increases monotonically and steeply with $\mathscr{C}$, depending weakly on the adiabatic index, Γ_2, of the core EOS. For the nonspinning BH–NS binaries, we derive the power-law relation between $\mathscr{C}$ and $f_{\rm cut}$ for $Q = 2$ and $\Gamma_2 = 3$ as $f_{\rm cut} \propto \mathscr{C}^{3.9}$, in which the power index of $\mathscr{C}$ is significantly larger than 1.5 which is expected from the analysis of the mass-shedding limit. This implies that the dependence of $f_{\rm cut}$ on $\mathscr{C}$ is stronger than that for $f_{\rm shed}$, and indicates that the observation of $f_{\rm cut}$ will play a role for constraining the value of $\mathscr{C}$.
5. The cutoff frequency of the gravitational-wave spectrum is correlated with the NS compactness in a clear manner when the NS is disrupted, and the BH spin modifies this correlation. The prograde BH spin decreases the cutoff frequency for fixed values of $\mathscr{C}$ and Q, because the angular velocity at the tidal disruption becomes smaller than that for $a = 0$. The cutoff frequency is lower for a smaller value of $\mathscr{C}$ for fixed values of Q and a, as in the case of nonspinning BH–NS binaries, because the tidal effect is stronger and the disruption occurs at a more distant orbit. The BH spin also modifies the spectrum for the inspiral phase. Specifically, the spectrum amplitude for a given frequency in the inspiral phase becomes large when the BH has a prograde spin, and this is consistent with the PN estimation. Both the low cutoff frequency and large spectrum amplitude in the inspiral phase for a prograde BH spin are encouraging for gravitational-wave astronomy to become a tool to investigate the NS compactness and EOS. It is noteworthy that the BH–NS binary with a high mass ratio of $Q \gtrsim 5$ is a more promising target for ground-based gravitational-wave detectors if the BH has a prograde spin and the NS tidal disruption occurs.
6. Varying the core EOS modifies the value of $f_{\rm cut}$, because the central density profile and tidal deformability of the NS depends on the stiffness of the core EOS. For the variation from $\Gamma_2 = 3$ to 2.4, the value of $f_{\rm cut}$ is modified by $\sim$20 %. This suggests that the details of the core EOS for $\rho \gtrsim 10^{15}\ {\rm g\,cm^{-3}}$ may play an important role for determining the gravitational waveform from the BH-NS binaries composed of high-mass NSs.

8.2 Future Work

Finally, we list several issues to be explored in the future.

1. The models of BH–NS binaries adopted in this thesis are not sufficiently astrophysically realistic because of a relatively small mass ratio, Q, and the complete alignment of the BH spin angular momentum to the orbital angular momentum of a binary. It is necessary to compute gravitational waves from a binary with the large mass ratio and with the BH spinning in the direction not aligned with the orbital angular velocity [1, 2]. It is also important to perform simulations of (nearly-)extremely spinning BH–NS binaries. It should be noted that, however, it is difficult to compute initial data with such a large spin in the puncture framework [3]. This issue must be overcome in the near future [4].
2. Piecewise polytropic EOSs with two pieces employed in this thesis are not accurate enough to model high-mass NSs with large central density of $\rho_{\max} \gtrsim 10^{15}\,\mathrm{g\,cm^{-3}}$ [5]. More detailed (piecewise polytropic) EOSs are necessary to calculate gravitational waves from a BH–relatively massive NS binary merger for which the tidal deformation and disruption of the NS plays an important role, i.e., BH–NS binaries with moderately large BH spins of $a \gtrsim 0.75$. The (undesirable) effect of approximate treatment of thermal effects is also a matter of debate.
3. The implementation of detailed microphysics, such as a finite-temperature effect and a neutrino transport process, is essential even qualitatively to explore the evolution of the remnant BH–accretion disk system and to discuss the jet launch such that short-hard GRBs require. Recently, we perform fully general relativistic simulations of binary NS mergers incorporating a finite-temperature nucleonic and hyperonic EOS with and an approximate neutrino emission scheme [6, 7]. We plan to work on BH–NS binary mergers along these lines. The magnetic field may also be important for the investigation of short-hard GRBs [8, 9].

References

1. F. Foucart, M.D. Duez, L.E. Kidder, S.A. Teukolsky, Phys. Rev. D **83**, 024005 (2011)
2. F. Foucart, M.D. Duez, L.E. Kidder, M.A. Scheel, B. Szilágyi, S.A. Teukolsky, Phys. Rev. D **85**, 044015 (2012)
3. G. Lovelace, R. Owen, H.P. Pfeiffer, T. Chu, Phys. Rev. D **78**, 084017 (2008)
4. Y.T. Liu, Z.B. Etienne, S.L. Shapiro, Phys. Rev. D **80**, 121503(R) (2009)
5. J.S. Read, B.D. Lackey, B.J. Owen, J.L. Friedman, Phys. Rev. D **79**, 124032 (2009)
6. Y. Sekiguchi, K. Kiuchi, K. Kyutoku, M. Shibata, Phys. Rev. Lett. **107**, 051102 (2011)
7. Y. Sekiguchi, K. Kiuchi, K. Kyutoku, M. Shibata, Phys. Rev. Lett. **107**, 211101 (2011)
8. S. Chawla, M. Anderson, M. Besselman, L. Lehner, S.L. Liebling, P.M. Motl, D. Neilsen, Phys. Rev. Lett. **105**, 111101 (2010)
9. Z.B. Etienne, Y.T. Liu, V. Paschalidis, S.L. Shapiro, Phys. Rev. D **85**, 064029 (2012)

Appendix A
The Canonical Formulation of General Relativity

The meaning of the $3 + 1$ decomposition and the Arnowitt-Deser-Misner (ADM) integral become clear when general relativity is written in the canonical formulation [1]. In this Appendix, we review the canonical formulation of general relativity and introduce the ADM integral as a surface integral (see [2, 3] for the details). The spacetime is denoted by $\mathscr{M}$, a spacelike hypersurface is denoted by Σ, and the timelike tube foliated by two surfaces $\mathscr{S}$ is denoted by $\mathscr{B}$.

A.1 Lagrangian Formulation

The Lagrangian density of general relativity is given by

$$\mathscr{L}_{\mathrm{G}} = {}^4R\sqrt{-g}, \tag{A.1}$$

and the Einstein equations are obtained by extremizing the Einstein-Hilbert action,

$$S_{\mathrm{G}} = \frac{1}{16\pi}\int_{\mathscr{M}} {}^4R\sqrt{-g}\mathrm{d}^4x, \tag{A.2}$$

with respect to the variation of the metric $\delta g^{\mu\nu}$. Here, the equation of motion is invariant up to a constant rescaling of the action, and the constant $1/16\pi$ is chosen for later convenience.[1] Using relations

[1] Specifically, we take this value so that the value of the Hamiltonian agrees with the value of the energy in the Newtonian limit.

Springer Theses, DOI: 10.1007/978-4-431-54201-8,

$$\delta\left(\sqrt{-g}\right) = \frac{1}{2}\sqrt{-g}g^{\alpha\beta}\delta g_{\alpha\beta} = -\frac{1}{2}\sqrt{-g}g_{\alpha\beta}\delta g^{\alpha\beta}, \tag{A.3}$$

$$\delta^4 R_{\mu\nu} = \frac{1}{2}\Big(-g^{\alpha\beta}\nabla_\alpha\nabla_\beta\delta g_{\mu\nu} - g^{\alpha\beta}\nabla_\mu\nabla_\nu\delta g_{\alpha\beta} + g^{\alpha\beta}\nabla_\alpha\nabla_\mu\delta g_{\nu\beta} + g^{\alpha\beta}\nabla_\alpha\nabla_\nu\delta g_{\mu\beta}\Big), \tag{A.4}$$

the variation of $S_{\rm G}$ is written as

$$\delta S_{\rm G} = \frac{1}{16\pi}\int_{\mathscr{M}} G_{\mu\nu}\delta g^{\mu\nu}\sqrt{-g}{\rm d}^4x + \frac{1}{16\pi}\int_{\mathscr{M}} \nabla^\mu v_\mu\sqrt{-g}{\rm d}^4x, \tag{A.5}$$

$$v_\mu \equiv \nabla^\alpha\delta g_{\mu\alpha} - g^{\alpha\beta}\nabla_\mu\delta g_{\alpha\beta}. \tag{A.6}$$

The action of the matter is generally written as

$$S_{\rm M} = \alpha_{\rm M}\int_{\mathscr{M}} \mathscr{L}_{\rm M}{\rm d}^4x, \tag{A.7}$$

where $\alpha_{\rm M}$ and $\mathscr{L}$ are a coupling constant and the matter Lagrangian density, respectively. The energy-momentum tensor is defined by

$$T_{\mu\nu} \equiv -\frac{2\alpha_{\rm M}}{\sqrt{-g}}\frac{\delta\mathscr{L}_{\rm M}}{\delta g^{\mu\nu}}, \tag{A.8}$$

and therefore the variation of $S_{\rm M}$ is written as

$$\delta S_{\rm M} = \frac{1}{16\pi}\int_{\mathscr{M}} -8\pi T_{\mu\nu}\delta g^{\mu\nu}\sqrt{-g}{\rm d}^4x. \tag{A.9}$$

When the divergence term of (A.5) can be dropped using the Gauss' theorem, the variation of $S = S_{\rm G} + S_{\rm M}$ with respect to the spacetime metric gives the Einstein equations,

$$G_{\mu\nu} = 8\pi T_{\mu\nu}, \tag{A.10}$$

in the bulk spacetime. Because S_G have to be invariant under diffeomorphisms, the contracted Bianchi identity $\nabla_\alpha G^{\mu\alpha} = 0$ is derived when we write the variation of metric as $\delta g^{\mu\nu} = -\pounds_w g^{\mu\nu}$ for a generator of the diffeomorphism, w^μ. The diffeomorphism invariance of $S_{\rm M}$ derives the local energy-momentum conservation equation, $\nabla_\alpha T^{\mu\alpha} = 0$, if the equation of motion is satisfied.[2]

[2] It should be noted that the equation of motion for the matter field is obtained by the variation of S_M with respect to the matter field, although the local energy-momentum conservation sometimes gives the equation of motion. For example, the Lagrangian density of a Klein-Gordon scalar field

It is not obvious that we can always drop the divergence term $\nabla^\mu v_\mu$, and indeed we cannot do in general. Such a complexity arises due to a peculiar feature of the Einstein-Hilbert action, which contains the second derivative of the metric in the scalar curvature. For a spacetime in which the value of the metric is fixed on the boundary $\partial\mathscr{M}$ but that of the first derivative is not, this action have to be supplemented with proper boundary terms [4, 5], i.e., the so-called Gibbons-Hawking-York term. For simplicity, we assume that the spacetime is bounded by two spacelike hypersurfaces Σ_1 (past) and Σ_2 (future) and a timelike tube foliated by two surfaces $\mathscr{B}$. The divergence term is rewritten using the Gauss' theorem and relations for a spacelike hypersurface,

$$n^\mu v_\mu = n^\mu \gamma^{\alpha\beta}(\nabla_\beta \delta g_{\mu\alpha} - \nabla_\mu \delta g_{\alpha\beta}) = -n^\mu \gamma^{\alpha\beta} \nabla_\mu \delta g_{\alpha\beta}, \tag{A.14}$$

$$\delta K = -\gamma_\mu{}^\alpha \delta C^\mu_{\alpha\beta} n^\beta = -\frac{1}{2} n^\mu \gamma^{\alpha\beta} \nabla_\mu \delta g_{\alpha\beta}, \tag{A.15}$$

where $\delta C^\mu_{\alpha\beta}$ is the difference between covariant derivatives (or the Christoffel symbol) associated with the original and perturbed metric. We also adopt similar relations for the timelike tube by denoting the induced metric and the extrinsic curvature[3] by

$$h_{\mu\nu} \equiv g_{\mu\nu} - s_\mu s_\nu, \tag{A.16}$$

$$\mathscr{K}_{\mu\nu} \equiv -\frac{1}{2} \pounds_s h_{\mu\nu}, \tag{A.17}$$

and we observe that the variation of the Einstein-Hilbert action leads

---

footnote 2 (continued)
ϕ with a given potential $V(\phi)$ (such as $m^2\phi^2$) is

$$\mathscr{L}_{\mathrm{KG}} = -\left[\frac{1}{2} g^{\alpha\beta} (\nabla_\alpha \phi)(\nabla_\beta \phi) + V(\phi)\right] \sqrt{-g}, \tag{A.11}$$

and hence the energy-momentum tensor is given by

$$T^{\mathrm{KG}}_{\mu\nu} = (\nabla_\mu \phi)(\nabla_\nu \phi) - g_{\mu\nu} \left[(\nabla_\alpha \phi)(\nabla^\alpha \phi) + V(\phi)\right], \tag{A.12}$$

where α_{KG} is chosen to be the unity. The equation of motion is given by the variation with respect to ϕ, and becomes

$$\frac{\delta \mathscr{L}_{\mathrm{KG}}}{\delta \phi} \propto \nabla^\alpha \nabla_\alpha \phi - V'(\phi) = 0, \tag{A.13}$$

where the prime denotes the derivative with respect to ϕ, and it is seen that this equation is also derived from $\nabla_\alpha T^{\mu\alpha} = 0$. However, we have to rely on the variation with respect to the fields when we have more than one scalar fields.

[3] The signature of the extrinsic curvature is not universal, and [2, 3] use opposite convention. Ours is consistent with, e.g., that of [6, 7].

$$\delta S_{\rm G} = \frac{1}{16\pi}\int_{\mathscr{M}} G_{\mu\nu}\delta g^{\mu\nu}\sqrt{-g}\mathrm{d}^4x + \frac{1}{8\pi}\int_{\mathscr{B}} \delta\mathscr{K}\sqrt{-h}\mathrm{d}^3x$$
$$-\frac{1}{8\pi}\int_{\Sigma_2}\delta K\sqrt{\gamma}\mathrm{d}^3x + \frac{1}{8\pi}\int_{\Sigma_1}\delta K\sqrt{\gamma}\mathrm{d}^3x. \quad \text{(A.18)}$$

Notice again that the value of the metric itself is fixed on the boundaries. This expression elucidates the necessity of adding Gibbons-Hawking-York terms,

$$S_{\rm GHY} = -\frac{1}{8\pi}\int_{\mathscr{B}}\mathscr{K}\sqrt{-h}\mathrm{d}^3x + \frac{1}{8\pi}\int_{\Sigma_2}K\sqrt{\gamma}\mathrm{d}^3x - \frac{1}{8\pi}\int_{\Sigma_1}K\sqrt{\gamma}\mathrm{d}^3x, \quad \text{(A.19)}$$

to have a well-posed variational problem. In order to have a finite value of the action, we have to also add

$$S_{\rm GHY,0} = \frac{1}{8\pi}\int_{\mathscr{B}}\mathscr{K}_0\sqrt{-h}\mathrm{d}^3x - \frac{1}{8\pi}\int_{\Sigma_2}K_0\sqrt{\gamma}\mathrm{d}^3x + \frac{1}{8\pi}\int_{\Sigma_1}K_0\sqrt{\gamma}\mathrm{d}^3x, \quad \text{(A.20)}$$

where $\mathscr{K}_0$ and K_0 are extrinsic curvatures evaluated at the flat spacetime. This term is not subjected to the variation of the metric. By adding all terms, we obtain the total action,

$$S = S_{\rm G} + S_{\rm M} + S_{\rm GHY} + S_{\rm GHY,0}. \quad \text{(A.21)}$$

A.2 Hamiltonian Formulation

Hamiltonian formulation of general relativity is derived in a usual way from the Lagrangian formulation with the aid of the $3+1$ decomposition. Hereafter, we always assume that the spacetime is foliated by a one-parameter family of spacelike hypersurfaces $\{\Sigma_t\}$, and denote the future-directed normal vector by n^μ. We also assume that the time vector is written by $t^\mu = \alpha n^\mu + \beta^\mu$. The $3+1$ decomposition of the gravitational part is performed using the relation,

$$^4R = {}^4R_{\mu\alpha\nu\beta}\gamma^{\mu\nu}\gamma^{\alpha\beta} - 2\,{}^4R_{\mu\nu}n^\mu n^\nu$$
$$= R - K^2 + K_{ij}K^{ij} + 2\nabla_\mu(n^\mu\nabla_\nu n^\nu - n^\nu\nabla_\nu n^\mu). \quad \text{(A.22)}$$

The first three terms give the bulk contribution, and the last divergence term gives the surface contribution. The surface term cancels with the Gibbons-Hawking-York term at Σ_1 and Σ_2 as is naturally expected, and only the integral at $\mathscr{B}$ contributes to the action. Combining with the Gibbons-Hawking-York term, we have

$$
\begin{aligned}
S_{\mathrm{G}}+S_{\mathrm{GHY}}+S_{\mathrm{GHY},0} = \frac{1}{16\pi}\int\Bigg[&\int_{\Sigma_t}(R-K^2+K_{ij}K^{ij})\alpha\sqrt{\gamma}\mathrm{d}^3x \\
&-2\oint_{\mathscr{S}_t}(k-k_0)\alpha\sqrt{q}\mathrm{d}^2x\Bigg]\mathrm{d}t,
\end{aligned}
\tag{A.23}
$$

where k_{AB} is the extrinsic curvature of the induced metric q_{AB} at a two surface, $\mathscr{S}_t$, and k is its trace. The trace of the extrinsic curvature of $\mathscr{S}_t$ evaluated at the flat spacetime is denoted by k_0.

The Hamiltonian density is derived from the bulk term, when we specify configuration variables. It is convenient to choose the configuration variables, which represents the spacetime metric, to be $(\alpha, \beta_i, \gamma_{ij})$. Recall the extrinsic curvature is related to these variables by

$$
K_{ij} = -\frac{1}{2\alpha}(\dot{\gamma}_{ij} - D_i\beta_j - D_j\beta_i), \tag{A.24}
$$

where the dot denotes the time derivative. The Lagrangian density is easily rewritten by inserting this relation, and the canonical conjugate momentum, or momentum density, variables are found to be

$$
\pi^0 \equiv \frac{\delta\mathscr{L}_{\mathrm{G}}}{\delta\dot{\alpha}} = 0, \tag{A.25}
$$

$$
\pi^i \equiv \frac{\delta\mathscr{L}_{\mathrm{G}}}{\delta\dot{\beta}_i} = 0, \tag{A.26}
$$

$$
\pi^{ij} \equiv \frac{\delta\mathscr{L}_{\mathrm{G}}}{\delta\dot{\gamma}_{ij}} = -\sqrt{\gamma}\left(K^{ij} - K\gamma^{ij}\right). \tag{A.27}
$$

It is important to observe that the momentum variables conjugate to (α, β_i) are not defined, because these configuration variables serve as gauge variables. The bulk Hamiltonian density is given by the Legendre transformation as

$$
\begin{aligned}
\mathscr{H}_{\mathrm{G}} &= \pi^{ij}\dot{\gamma}_{ij} - \mathscr{L}_{\mathrm{G}} \\
&= -\alpha\sqrt{\gamma}R + \frac{\alpha}{\sqrt{\gamma}}\left(\pi^{ij}\pi_{ij} - \frac{1}{2}\pi^2\right) \\
&\quad + 2\sqrt{\gamma}D_i\left(\frac{\pi^{ij}\beta_j}{\sqrt{\gamma}}\right) - 2\sqrt{\gamma}\beta_j D_i\left(\frac{\pi^{ij}}{\sqrt{\gamma}}\right)
\end{aligned}
\tag{A.28}
$$

$$
\begin{aligned}
&= -\alpha\sqrt{\gamma}\left(R + K^2 - K_{ij}K^{ij}\right) - 2\sqrt{\gamma}D_i\left[\beta_j(K^{ij} - K\gamma^{ij})\right] \\
&\quad + 2\sqrt{\gamma}\beta_j D_i(K^{ij} - K\gamma^{ij}),
\end{aligned}
\tag{A.29}
$$

and the Hamiltonian is given by an integration on Σ_t as

$$H_{\mathrm{G}} = \frac{1}{16\pi}\int_{\Sigma_t}\left[-\alpha\left(R+\frac{1}{2\gamma}\pi^2-\frac{1}{\gamma}\pi^{ij}\pi_{ij}\right)-2\beta_j D_i\left(\frac{\pi^{ij}}{\sqrt{\gamma}}\right)\right]\sqrt{\gamma}\mathrm{d}^3x$$
$$+\frac{1}{8\pi}\oint_{\mathscr{S}_t}\left[\alpha(k-k_0)+\frac{\beta_j\pi^{ij}s_i}{\sqrt{\gamma}}\right]\sqrt{q}\mathrm{d}^2x \tag{A.30}$$
$$=\frac{1}{16\pi}\int_{\Sigma_t}\left[-\alpha\left(R+K^2-K_{ij}K^{ij}\right)+2\beta_j D_i\left(K^{ij}-K\gamma^{ij}\right)\right]\sqrt{\gamma}\mathrm{d}^3x$$
$$+\frac{1}{8\pi}\oint_{\mathscr{S}_t}\left[\alpha(k-k_0)-\beta_j\left(K^{ij}-K\gamma^{ij}\right)s_i\right]\sqrt{q}\mathrm{d}^2x. \tag{A.31}$$

The Hamiltonian density of the matter is simply given by

$$\mathscr{H}_{\mathrm{M}} = -16\pi\alpha_{\mathrm{M}}\mathscr{L}_{\mathrm{M}}, \tag{A.32}$$

because the Lagrangian density of the matter does not depend on the time derivative of the metric quantities as far as we concern. Here, we include $16\pi\alpha_{\mathrm{M}}$ into the definition of the Hamiltonian density for simplicity, and the total Hamiltonian density is given by the sum of gravitational and matter terms.

Before going to the variation of the gravitational term, we derive the expression for the variation of the matter term. Using the fact that the spacetime metric is written as

$$g^{\mu\nu} = \gamma^{\mu\nu} - \frac{1}{\alpha^2}(t^\mu-\beta^\mu)(t^\nu-\beta^\nu), \tag{A.33}$$

the variation of $g^{\mu\nu}$ with respect to $(\alpha, \beta_i, \gamma_{ij})$ are given by

$$\frac{\delta g^{\mu\nu}}{\delta\alpha} = \frac{2}{\alpha}n^\mu n^\nu, \tag{A.34}$$
$$\frac{\delta g^{\mu\nu}}{\delta\beta_i} = \frac{2}{\alpha}\gamma^{i\mu}n^\nu, \tag{A.35}$$
$$\frac{\delta g^{\mu\nu}}{\delta\gamma_{ij}} = -\gamma^{i\mu}\gamma^{j\nu} - \frac{2}{\alpha}\gamma^{i\mu}n^\nu\beta^j. \tag{A.36}$$

Therefore, the variation of the matter Hamiltonian density is found to be

$$\frac{\delta\mathscr{H}_{\mathrm{M}}}{\delta\alpha} = 16\pi\sqrt{\gamma}\rho_{\mathrm{H}}, \tag{A.37}$$
$$\frac{\delta\mathscr{H}_{\mathrm{M}}}{\delta\beta_i} = -16\pi\sqrt{\gamma}j^i, \tag{A.38}$$
$$\frac{\delta\mathscr{H}_{\mathrm{M}}}{\delta\gamma_{ij}} = -8\pi\alpha\sqrt{\gamma}S^{ij} + 16\pi\sqrt{\gamma}j^i\beta^j, \tag{A.39}$$

where we used (A.8) and the definition of variables in Sect. 3.2.1, (3.7), (3.8), and (3.9).

The variation of the Hamiltonian density with respect to the lapse function, shift vector, induced metric, and the momentum density derives the Hamiltonian constraint, momentum constraint, evolution equation of the momentum density, and evolution equation of the induced metric, respectively. Here, the boundary condition is taken to be

$$\delta\alpha = 0, \delta\beta_i = 0, \delta\gamma_{ij} = 0, \tag{A.40}$$

and $\delta\pi^{ij}$ is not constrained. First, the variation with respect to (α, β_i) have to be zero, so that (π^0, π^i) do not evolve in time. Because the surface term is not varied due to the boundary condition, we obtain the Hamiltonian constraint, (3.10), and the momentum constraint, (3.11). To vary the Hamiltonian density with respect to (γ_{ij}, π^{ij}), it is useful to rewrite the bulk term of the gravitational part as

$$\mathscr{H}_\mathrm{G} = -\alpha\sqrt{\gamma}R + \frac{\alpha}{\sqrt{\gamma}}\left(\pi^{ij}\pi_{ij} - \frac{1}{2}\pi^2\right) + \pi^{ij}D_i\beta_j + \pi^{ij}D_j\beta_i. \tag{A.41}$$

The evolution equation of the induced metric is given by a standard relation,

$$\dot{\gamma}_{ij} = \frac{\delta\mathscr{H}}{\delta\pi^{ij}}, \tag{A.42}$$

and this gives (3.12). Note that this relation involves no matter terms, because the matter Hamiltonian does not depend on π^{ij}. The evolution equation of the momentum density is given by

$$\dot{\pi}^{ij} = -\frac{\delta\mathscr{H}}{\delta\gamma_{ij}}, \tag{A.43}$$

and this variation is accomplished using relations derived in the previous section, say δR_{ij} and δC^k_{ij}. It is customary to use the momentum constraint to derive this evolution equation. It should be noted that the surface contribution derived from the variation of the bulk Hamiltonian cancels with the variation of the surface Hamiltonian, as is naturally expected. The evolution equation of the extrinsic curvature, (3.13), is obtained by the definition of π^{ij}, evolution equations, and the Hamiltonian constraint.

A.3 Boundary Values and the ADM Integral

The value of the Hamiltonian in an instant is solely determined by the boundary terms,

$$H = \frac{1}{8\pi}\oint_{\mathscr{S}_t}\left[\alpha(k - k_0) - \beta_i(K^{ij} - K\gamma^{ij})s_j\right]\sqrt{q}\mathrm{d}^2x, \tag{A.44}$$

when the Hamiltonian and momentum constraints are satisfied. The energy and momentum of the system are defined via this boundary value of the Hamiltonian [1, 8].

Because the energy is the charge associated with the time translation, the value of the ADM energy is computed by an inertial observer, ($\alpha = 1, \beta_i = 0$), at spatial infinity by [9]

$$E_{\mathrm{ADM}} \equiv \frac{1}{8\pi} \lim_{\mathscr{S}_t \to \infty} \oint_{\mathscr{S}_t} (k - k_0)\sqrt{q} \mathrm{d}^2 x. \tag{A.45}$$

By rewriting this [10], an original expression of the ADM energy,

$$E_{\mathrm{ADM}} = \frac{1}{16\pi} \lim_{\mathscr{S}_t \to \infty} \oint_{\mathscr{S}_t} \left[\mathring{D}^j \gamma_{ij} - \mathring{D}_i (f^{jk} \gamma_{jk})\right] s^i \sqrt{q} \mathrm{d}^2 x, \tag{A.46}$$

is obtained when we assume sufficiently rapid falloff of the induced metric, i.e., asymptotic flatness condition. Specifically,

$$\gamma_{ij} - f_{ij} = O(r^{-1}), \ \partial_k \gamma_{ij} = O(r^{-2}), \tag{A.47}$$

in Cartesian-type coordinates. This expression is the same as (3.60), where we denote this quantity as the ADM mass, M_{ADM}.

The ADM linear momentum is defined using asymptotic translational Killing vectors, $\partial^j_{(i)}$, where the subscript $i = x, y, z$ denotes the label of the Killing vector. The ADM linear momentum is obtained by setting $\alpha = 0$ and $\beta_j = \partial_{(i)j}$ as

$$P_i \equiv \frac{1}{8\pi} \lim_{\mathscr{S}_t \to \infty} \oint_{\mathscr{S}_t} (K_{jk} - K\gamma_{jk}) \partial^j_{(i)} s^k \sqrt{q} \mathrm{d}^2 x, \tag{A.48}$$

where we also assume sufficiently rapid falloff of the extrinsic curvature,

$$K_{ij} = O(r^{-2}), \ \partial_k K_{ij} = O(r^{-3}). \tag{A.49}$$

The expression above agrees with (3.64) when we substitute $\partial^j_{(i)} = \delta_i{}^j$. The ADM energy and linear momentum composes an ADM 4-momentum vector,

$$P_\mu = (-E_{\mathrm{ADM}}, P_i), \tag{A.50}$$

which behaves in a proper way under the Poincaré transformation [1].

An ADM-like angular momentum may be defined using asymptotic rotational Killing vectors, $\phi^j_{(i)}$, as

$$J_i \equiv \frac{1}{8\pi} \lim_{\mathscr{S}_t \to \infty} \oint_{\mathscr{S}_t} (K_{jk} - K\gamma_{jk}) \phi^j_{(i)} x s^k \sqrt{q} \mathrm{d}^2 x. \tag{A.51}$$

Using the fact that the rotational Killing vector is expressed as $\phi_{(i)j} = \overset{\circ}{\varepsilon}_{jkl}\delta_i{}^k x^l$ and is orthogonal to s^i, it is shown that this expression is equivalent to (3.66). However, the care must be taken to compute this quantity, because the value of the ADM-like angular momentum is known to depend on the gauge choice via the so-called supertranslation ambiguity. To overcome this, we should adopt the quasi-isotropic and asymptotically maximal gauge [6].

Appendix B
Gravitational Waves

Gravitational waves are no doubt important for the purpose of this thesis. In this Appendix, we first review basic properties of gravitational waves [11]. Next, we summarize the radiation of the energy, linear momentum, and the angular momentum, focusing on the decomposition using spin-weighted spherical harmonics (see e.g., [12] for the details). In this Appendix, the partial derivative and the covariant derivative associated with $\eta_{\mu\nu}$ are both denoted by ∂_μ, and we expect that the meaning can be understood in the context.

B.1 The Propagation of Gravitational Waves

Gravitational waves are identified as the perturbation on the background metric, which we typically consider to be the flat one. For later convenience, we consider the background metric $\bar{g}_{\mu\nu}$ as a general metric, and investigate the propagation of the perturbation $h_{\mu\nu}$ on the background. We assume that the perturbation is sufficiently small in the sense that the condition $|h_{\mu\nu}| \leq 1$ is satisfied in Cartesian-like coordinates. Hereafter, typical amplitude of the perturbation is denoted by h, although we also use h for its trace, $h \equiv \bar{g}^{\mu\nu} h_{\mu\nu}$. The full spacetime metric is written by

$$g_{\mu\nu} = \bar{g}_{\mu\nu} + h_{\mu\nu}, \tag{B.1}$$

and its inverse is written to a linear order of h by

$$g^{\mu\nu} = \bar{g}^{\mu\nu} - h^{\mu\nu} + O(h^2), \tag{B.2}$$

where indices of $h_{\mu\nu}$ is raised and lowered by the background metric, $\bar{g}_{\mu\nu}$. The first order deviation of the Christoffel symbol and Riemann tensor from the background ones, which are computed solely by $\bar{g}^{\mu\nu}$, are written as

$${}^{(1)}\Gamma^\alpha{}_{\mu\nu} = \frac{1}{2}\bar{g}^{\alpha\beta}\left(\nabla_\mu h_{\nu\beta} + \nabla_\nu h_{\mu\beta} - \nabla_\beta h_{\mu\nu}\right), \tag{B.3}$$

K. Kyutoku, *The Black Hole–Neutron Star Binary Merger in Full General Relativity*, Springer Theses, DOI: 10.1007/978-4-431-54201-8, © Springer Japan 2013

$$^{(1)}R_{\alpha\mu\beta}{}^{\nu} = \frac{1}{2}\Big(\nabla_\mu\nabla_\alpha h_\beta{}^\nu + \nabla_\mu\nabla_\beta h_\alpha{}^\nu - \nabla_\mu\nabla^\nu h_{\alpha\beta} - \nabla_\alpha\nabla_\mu h_\beta{}^\nu - \nabla_\alpha\nabla_\beta h_\mu{}^\nu + \nabla_\alpha\nabla^\nu h_{\beta\mu}\Big), \tag{B.4}$$

where ∇_μ denotes the covariant derivative associated with the background metric, $\bar{g}_{\mu\nu}$, in this Appendix. The first-order Ricci tensor is

$$^{(1)}R_{\mu\nu} = \frac{1}{2}\Big(-\nabla^\alpha\nabla_\alpha h_{\mu\nu} - \nabla_\mu\nabla_\nu h + \nabla_\mu\nabla^\alpha h_{\alpha\nu} + \nabla_\nu\nabla^\alpha h_{\alpha\mu} + \bar{R}_{\mu\alpha}h_\nu{}^\alpha + \bar{R}_{\nu\alpha}h_\mu{}^\alpha + \bar{R}_{\alpha\mu\nu\beta}h^{\alpha\beta} + \bar{R}_{\alpha\nu\mu\beta}h^{\alpha\beta}\Big), \tag{B.5}$$

where barred quantities in this expression denote the background curvature, and the scalar curvature is

$$^{(1)}R = \bar{g}^{\mu\nu}\,{}^{(1)}R_{\mu\nu} - h^{\mu\nu}\bar{R}_{\mu\nu}. \tag{B.6}$$

The Einstein tensor is computed straightforwardly by these quantities, and it is more conveniently expressed using the trace-reverse tensor defined by

$$\bar{h}_{\mu\nu} \equiv h_{\mu\nu} - \frac{1}{2}h\bar{g}_{\mu\nu}, \tag{B.7}$$

as

$$^{(1)}G_{\mu\nu} = \frac{1}{2}\Big(-\nabla^\alpha\nabla_\alpha\bar{h}_{\mu\nu} + \nabla_\mu\nabla^\alpha\bar{h}_{\nu\alpha} + \nabla_\nu\nabla^\alpha\bar{h}_{\mu\alpha} - \bar{g}_{\mu\nu}\nabla^\alpha\nabla^\beta\bar{h}_{\alpha\beta} + \bar{R}_{\mu\alpha}\bar{h}_\nu{}^\alpha + \bar{R}_{\nu\alpha}\bar{h}_\mu{}^\beta - 2\bar{R}_{\mu\alpha\nu\beta}\bar{h}^{\alpha\beta} - \bar{h}_{\mu\nu}\bar{R} + \bar{g}_{\mu\nu}\bar{h}^{\alpha\beta}\bar{R}_{\alpha\beta}\Big). \tag{B.8}$$

Hereafter in this section, we only consider the vacuum spacetime, and hence both the energy-momentum tensor and background Ricci tensor vanish. The first-order Einstein equations derive the equation governing $\bar{h}_{\mu\nu}$ as

$$\nabla^\alpha\nabla_\alpha\bar{h}_{\mu\nu} + \bar{g}_{\mu\nu}\nabla^\alpha\nabla^\beta\bar{h}_{\alpha\beta} - \nabla_\mu\nabla^\alpha\bar{h}_{\nu\alpha} - \nabla_\nu\nabla^\alpha\bar{h}_{\mu\alpha} = -2\bar{R}_{\mu\alpha\nu\beta}\bar{h}^{\alpha\beta}. \tag{B.9}$$

The discussion becomes clear when the harmonic gauge,[4]

---

[4] This condition is equivalent to $\nabla^\alpha\nabla_\alpha x^\mu = 0$ to a linear order in $h_{\mu\nu}$. In the post-Newtonian expansion, the background metric is chosen to be the flat one and the harmonic coordinate is defined by

$$\partial_\alpha\left(\sqrt{-g}g^{\mu\alpha} - \eta^{\mu\alpha}\right) = 0. \tag{B.10}$$

$$\nabla_\alpha \bar{h}^{\mu\alpha} = 0, \tag{B.11}$$

is chosen. In this gauge, the first-order Einstein equations in a vacuum become

$$\nabla^\alpha \nabla_\alpha \bar{h}_{\mu\nu} = -2\bar{R}_{\mu\alpha\nu\beta}\bar{h}^{\alpha\beta}. \tag{B.12}$$

In particular, when the background spacetime is the flat spacetime, the first-order Einstein equations become

$$\partial^\alpha \partial_\alpha \bar{h}_{\mu\nu} = 0, \tag{B.13}$$

which clearly shows that the perturbation of the metric propagates as waves, i.e., gravitational waves. The residual gauge freedom[5] of the harmonic gauge is fixed by taking the transverse-traceless gauge,

$$\bar{h} = \bar{h}^{ti} = 0, \tag{B.14}$$

and the time component of the harmonic gauge further derives $\dot{\bar{h}}^{tt} = 0$, if we have $\bar{h}^{tt} = 0$ is satisfied at a certain instant. The traceless nature implies that $h_{\mu\nu}$ and $\bar{h}_{\mu\nu}$ are identical in this gauge. The spatial components of the harmonic gauge give

$$\partial_j h^{ij} = 0, \tag{B.15}$$

and this equation combined with the fact that $h_{\mu\nu}$ satisfies the wave equation shows the transverse nature of $h_{\mu\nu}$. Specifically, when the direction of propagation is chosen to be the z axis with the frequency ω and the wave vector $k^i = (0, 0, k)$, gravitational waves in the transverse-traceless gauge is written by two independent functions h_+ and $h_\times$ as

$$h_{\mu\nu} = h_+(t,z)e^+_{\mu\nu} + h_\times(t,z)e^\times_{\mu\nu}, \tag{B.16}$$

$$h_+(t,z) = A_+\cos(\omega t - kz + \Phi_+)\,,\ h_\times(t,z) = A_\times\cos(\omega t - kz + \Phi_\times), \tag{B.17}$$

$$\begin{aligned} e^+_{\mu\nu} &= \left[(e_x)_\mu(e_x)_\nu - (e_y)_\mu(e_y)_\nu\right] \\ e^\times_{\mu\nu} &= \left[(e_x)_\mu(e_y)_\nu + (e_y)_\mu(e_x)_\nu\right], \end{aligned} \tag{B.18}$$

where $(e_x)^\mu$ and $(e_y)^\mu$ are orthonormal bases in the x and y direction, respectively. The combination of them, $e^+_{\mu\nu}$ and $e^\times_{\mu\nu}$ are called the polarization tensors. The amplitude of each mode, A_+ and $A_\times$, are arbitrary, and the phase, Φ_+ and $\Phi_\times$, are determined by the initial condition. Because the first order equation is a linear equation, general gravitational waves are described by a superposition of this wave solution. The physical, gauge-invariant information of gravitational waves are totally described by

[5] The harmonic gauge gives the evolution of gauge variables, i.e., the lapse function and shift vector, in the 3 + 1 language. Fixing the residual corresponds to giving initial data of the lapse function and shift vector.

these two functions, h_+ and $h_\times$, and they are written as

$$h_+ = \frac{1}{2}\left(h_{\hat{\theta}\hat{\theta}} - h_{\hat{\varphi}\hat{\varphi}}\right), \tag{B.19}$$

$$h_\times = h_{\hat{\theta}\hat{\varphi}} = h_{\hat{\varphi}\hat{\theta}}, \tag{B.20}$$

in the orthonormal basis of spherical coordinates. Assuming that gravitational waves are propagating in the radial direction, which have the form $\sim h(t-r)/r$ at a large separation from the origin, i.e., $\partial_r h = -\dot{h}$, the component of the Riemann tensor is related to the metric perturbation by

$$R_{titj} = -R_{tirj} = R_{rirj} = -\frac{1}{2}\frac{\partial^2 h_{ij}}{\partial t^2}. \tag{B.21}$$

It is also important that the first-order Riemann tensor around the Minkowski spacetime is gauge-invariant quantity, and therefore the Riemann tensor computed in the transverse-traceless gauge has physical meanings.

B.2 The Energy-Momentum Tensor of Gravitational Waves

The energy-momentum tensor of gravitational waves (or gravitons) are not defined locally, because it is always possible to eliminate them due to the equivalence principle. Instead, an averaged energy-momentum tensor can be defined via the second-order perturbation of the Ricci tensor [13, 14]. The other method to define the energy-momentum tensor, again as an averaged quantity, is to rely on the Noether's theorem.

First, we derive the energy-momentum tensor from the metric perturbation. We assume that the "background metric" and "gravitational waves" is always distinguished from their typical length and/or time scales [15]. The background is identified as the component with large length and/or long time scales, and gravitational waves are identified as the component with small length and/or short time scales. For simplicity, we focus on the case in which the background has a typical frequency f_{B}, which is much smaller than the frequency of gravitational waves, f. The Ricci tensor is expanded by the order of the metric perturbation, h, around the background value as

$$R_{\mu\nu} = \bar{R}_{\mu\nu} + {}^{(1)}R_{\mu\nu} + {}^{(2)}R_{\mu\nu} + O(h^3), \tag{B.22}$$

and decomposition of components yields

$$\bar{R}_{\mu\nu} = -\left({}^{(2)}R_{\mu\nu}\right)^{\text{low}} + 8\pi\left(T_{\mu\nu} - \frac{1}{2}g_{\mu\nu}T\right)^{\text{low}}, \tag{B.23}$$

for low frequency, and

$${}^{(1)}R_{\mu\nu} = -\left({}^{(2)}R_{\mu\nu}\right)^{\text{high}} + 8\pi\left(T_{\mu\nu} - \frac{1}{2}g_{\mu\nu}T\right)^{\text{high}}, \tag{B.24}$$

for high frequency. The second-order perturbation of the Ricci tensor is given by

$$\begin{aligned}
{}^{(2)}R_{\mu\nu} = \frac{1}{2}\bar{g}^{\alpha\beta}\bar{g}^{\lambda\sigma}\Big[& h_{\alpha\lambda}(\nabla_\nu\nabla_\mu h_{\beta\sigma} + \nabla_\beta\nabla_\sigma h_{\mu\nu} - \nabla_\beta\nabla_\nu h_{\mu\sigma} - \nabla_\beta\nabla_\mu h_{\nu\sigma}) \\
& + \frac{1}{2}\nabla_\mu h_{\alpha\lambda}\nabla_\nu h_{\beta\sigma} + \nabla_\lambda h_{\nu\alpha}(\nabla_\sigma h_{\mu\beta} - \nabla_\beta h_{\mu\sigma}) \\
& + \left(\frac{1}{2}\nabla_\alpha h_{\lambda\sigma} - \nabla_\lambda h_{\alpha\sigma}\right) \\
& \times (\nabla_\nu h_{\mu\beta} + \nabla_\mu h_{\nu\beta} - \nabla_\beta h_{\mu\nu})\Big],
\end{aligned} \tag{B.25}$$

where we dropped background curvature terms associated with the permutation of derivatives, because we assume that the background curvature is associated with large scale and/or slowly varying components of the background metric.

The r.h.s of the low-frequency equation, (B.23), gives the definition of the "energy-momentum tensor of gravitational waves" as a source of the spacetime curvature due to the terms quadratic in gravitational waves, $h_{\mu\nu}$. It should be noted that, in the absence of the matter, the smallness of $\bar{R}_{\mu\nu}$ and ${}^{(2)}R_{\mu\nu}$ has to balance for this equation to hold, and therefore we find that $h \sim f_{\rm B}/f$ holds.[6] Because the separation is done in the intermediate scale between $f_{\rm B}$ and f, gravitational waves have to be averaged over several periods or wavelengths. Because these two are equivalent for gravitational waves, we perform spatial averaging of ${}^{(2)}R_{\mu\nu}$, and therefore the gravitational-wave energy-momentum tensor is defined by

$$t_{\mu\nu} \equiv -\frac{c^4}{8\pi G}\left\langle {}^{(2)}R_{\mu\nu} - \frac{1}{2}\bar{g}_{\mu\nu}{}^{(2)}R\right\rangle, \tag{B.26}$$

where the bracket denotes the spatial average and the trace of second-order Ricci tensor is taken with respect to the background metric. We inserted G and c for clarity. The averaging of the second-order Einstein tensor around the flat spacetime is now sufficient to derive an explicit expression, and we obtain

$$t_{\mu\nu} = \frac{c^4}{32\pi G}\langle \partial_\mu h_{ij}\partial_\nu h^{ij}\rangle, \tag{B.27}$$

[6] On the other hand, the high-frequency component shown that gravitational waves propagate on the curved background if it is a vacuum.

where we evaluate this in the transverse-traceless gauge, make use of the wave equation, and perform integration by parts to drop the surface terms. We inserted G and c for clarity. This quantity is invariant under the residual gauge transformation of the harmonic gauge, and therefore physically meaningful. In the flat and vacuum spacetime, the gravitational-wave energy-momentum tensor satisfies the conservation equation,

$$\partial_\alpha t^{\mu\alpha} = 0. \tag{B.28}$$

The energy contained in the volume is evaluated by the spatial integration of t^{tt}, and the linear momentum is evaluated by the integration of t^{ti}. The radiation flux is defined by the loss rate of these integrated quantities, and the use of the conservation equation and the Gauss' theorem derives the expression for the gravitational-wave flux. It is convenient to express these quantities in terms of a complex gravitational-wave function $H \equiv h_+ - ih_\times$, and the energy and linear momentum fluxes are written as

$$\frac{\mathrm{d}E}{\mathrm{d}t} = \lim_{r\to\infty} \frac{r^2}{16\pi} \oint_{\mathscr{S}} \left|\dot{H}\right|^2 \mathrm{d}\Omega, \tag{B.29}$$

$$\frac{\mathrm{d}P_i}{\mathrm{d}t} = \lim_{r\to\infty} \frac{r^2}{16\pi} \oint_{\mathscr{S}} s_i \left|\dot{H}\right|^2 \mathrm{d}\Omega. \tag{B.30}$$

Here we used the fact that the function H takes the form $\sim H(t - r)/r$ at spatial infinity, i.e., $\partial_r H = -\dot{H}$. Notice that this expression is also obtained by simply integrating the $t^{t\mu}$ on a coordinate sphere and taking the limit of infinite distance.

Deriving the angular momentum of gravitational waves is not straightforward with the energy-momentum tensor [16]. It is easier to compute the angular momentum by the Noether's theorem with respect to the rotational symmetry of the flat spacetime. The Lagrangian of gravitons, or gravitational waves, are derived by expanding the Einstein-Hilbert action to the second order around the flat spacetime, and it derives

$${}^{(2)}\mathscr{L}_{\mathrm{G}} = -\frac{1}{4\pi}\left(\partial_\mu h_{\alpha\beta}\partial^\mu h^{\alpha\beta} - \partial_\mu h \partial^\mu h + 2\partial_\mu h^{\mu\nu}\partial_\nu h - 2\partial_\mu h^{\mu\nu}\partial_\alpha h^\alpha_\nu\right). \tag{B.31}$$

The standard procedure to derive the energy-momentum tensor in the harmonic gauge gives the same result as (B.27), taking the prefactor $1/16\pi$ into account, as is naturally expected. The angular momentum of gravitational waves can be defined in the transverse-traceless gauge as a Noether charge with respect to the spatial rotation symmetry (see [11] for the detail) as

$$J^i = \frac{c^3}{32\pi G}\int \left(-\mathring{\varepsilon}^{ikl}\dot{h}^{mn} x_k \partial_l h_{mn} + 2\mathring{\varepsilon}^{ikl} h_k{}^m \dot{h}_{ml}\right) \mathrm{d}^3 x, \tag{B.32}$$

where we again take the prefactor into account and inserted G and c for clarity. The first term in the integrand is considered to be the orbital angular momentum, and the second term is the spin-2 angular momentum of the graviton. The angular momentum flux is evaluated by the integral of this charge behind the wave front, and a straightforward computation shows that [12]

$$\frac{\mathrm{d}J^i}{\mathrm{d}t} = -\lim_{r\to\infty} \frac{r^2}{32\pi} \oint_{\mathscr{S}} \dot{h}^{ij} \pounds_{\phi_{(i)}} h_{ij} \mathrm{d}\Omega, \tag{B.33}$$

where $\phi_{(i)}$ is the rotational Killing vector. Further, introducing two complex vectors

$$(\phi_\pm)^i \equiv (\phi_x)^i \pm i(\phi_y)^i, \tag{B.34}$$

and another two complex vectors[7]

$$(e_\pm)^i \equiv \frac{1}{\sqrt{2}} \left[(e_{\hat{\theta}})^i \mp i(e_{\hat{\varphi}})^i \right], \tag{B.35}$$

it is shown that gravitational waves in the transverse-traceless gauge, (B.16), is written as

$$h_{ij} = H(e_-)_i(e_-)_j + \bar{H}(e_+)_i(e_+)_j. \tag{B.36}$$

This expression shows that H and $\bar{H}$ has the spin weight[8] -2 and 2, respectively. We further define two angular momentum operators,

$$\hat{J}_\pm \equiv \phi^i_\pm \partial_i - \mathrm{i}\frac{s}{\sin\theta} \mathrm{e}^{\pm \mathrm{i}\varphi}, \tag{B.37}$$

and it is shown with some computation that

$$\dot{h}^{ij} \pounds_{\phi_\pm} h_{ij} = 2\mathrm{Re}\left[\dot{\bar{H}} \hat{J}_\pm H \right], \tag{B.38}$$

Finally, using the angular momentum operators,

---

[7] These correspond to two of null tetrads $\bar{m}^\mu$ and m^μ in the flat spacetime.

[8] A scalar f is called "it has a spin weight s" if it transforms as $f \to e^{-\mathrm{i}s\psi} f$ when the coordinate bases are rotated by an angle ψ. The spin weight of the component of tensorial quantities is easily understood by the number of e_+ and e_- attached to it.

$$\hat{J}_x \equiv \frac{1}{2}\left(\hat{J}_+ + \hat{J}_-\right)$$
$$= -\sin\varphi\partial_\theta - \cos\varphi\left(\cot\theta\partial_\varphi - \mathrm{i}\frac{s}{\sin\theta}\right), \tag{B.39}$$
$$\hat{J}_y \equiv \frac{1}{2\mathrm{i}}\left(\hat{J}_+ - \hat{J}_-\right)$$
$$= \cos\varphi\partial_\theta - \sin\varphi\left(\cot\theta\partial_\varphi - \mathrm{i}\frac{s}{\sin\theta}\right), \tag{B.40}$$
$$\hat{J}_z = \partial_\varphi, \tag{B.41}$$

the angular momentum flux is written as

$$\frac{\mathrm{d}J_i}{\mathrm{d}t} = -\lim_{r\to\infty}\frac{r^2}{16\pi}\mathrm{Re}\oint_{\mathscr{S}}\dot{\bar{H}}\hat{J}_i H \mathrm{d}\Omega. \tag{B.42}$$

B.3 The Mode Decomposition of the Flux

In numerical relativity, gravitational waves are often extracted by the Weyl scalar Ψ_4, which has a spin weight -2 and is related to H by $\Psi_4 = \ddot{H}$ (see Sect. 5.1). Because it is customary to decompose Ψ_4 using the spin-weighted spherical harmonics as

$$\Psi_4(r,\theta,\varphi) = \sum_{l=2}^{\infty}\sum_{m=-l}^{l}\Psi_4^{lm}(r)_{-2}Y_{lm}(\theta,\varphi), \tag{B.43}$$

it is quite useful to give the expression of the flux in terms of the coefficient for each mode (see [12] for the detail). The spin-weighted spherical harmonics is related to the complex conjugate by,

$${}_s\bar{Y}^{lm} = (-1)^{m+s}{}_{-s}Y^{l,-m}, \tag{B.44}$$

satisfy the orthogonality relation,

$$\oint_{\mathscr{S}} {}_sY^{lm}(\theta,\varphi){}_{s'}\bar{Y}^{l'm'}d\Omega = \delta_{ss'}\delta_{ll'}\delta_{mm'}, \tag{B.45}$$

and the integration of triple product are given by

$$\oint_{\mathscr{S}} {}_{s_1}Y^{l_1m_1}(\theta,\varphi){}_{s_2}Y^{l_2m_2}(\theta,\varphi){}_{s_3}Y^{l_3m_3}(\theta,\varphi)\mathrm{d}\Omega$$
$$= \sqrt{\frac{(2l_1+1)(2l_2+1)(2l_3+1)}{4\pi}}\begin{pmatrix} l_1 & l_2 & l_3 \\ -s_1 & -s_2 & -s_3 \end{pmatrix}\begin{pmatrix} l_1 & l_2 & l_3 \\ m_1 & m_2 & m_3 \end{pmatrix}, \tag{B.46}$$

where the Wigner 3j symbol are given by, for example,

$$\begin{pmatrix} l_1 & l_2 & l_3 \\ m_1 & m_2 & m_3 \end{pmatrix} = (-1)^{l_1-m_1}\delta_{m_1+m_2+m_3,0}$$
$$\times\sqrt{\frac{(l_1+l_2-l_3)!(l_3+l_1-l_2)!(l_2+l_3-l_1)!(l_3+m_3)!(l_3-m_3)!}{(l_1+l_2+l_3+1)!(l_1+m_1)!(l_1-m_1)!(l_2+m_2)!(l_2-m_2)!}}$$
$$\times\sum_{k\geq 0}\frac{(-1)^k}{k!}\frac{(l_2+l_3+m_1-k)!(l_1-m_1+k)!}{(l_3-l_1+l_2-k)!(l_3-m_3-k)!(l_1-l_2+m_3+k)!}. \tag{B.47}$$

In the Wigner 3j symbol, the sum is taken over as long as numbers in parentheses are all non-negative. The angular momentum operators act on the spin-weighted spherical harmonics as

$$\hat{J}_{zs}Y^{lm} = \mathrm{i}m_sY^{lm}, \tag{B.48}$$
$$\hat{J}_{\pm s}Y^{lm} = \mathrm{i}\sqrt{(l\mp m)(l+1\pm m)}{}_sY^{l,m\pm 1}, \tag{B.49}$$

and this is the same as the standard spherical harmonics apart from the definition of the angular momentum operator.

The energy flux is simply given by

$$\frac{\mathrm{d}E}{\mathrm{d}t} = \lim_{r\to\infty}\frac{r^2}{16\pi}\sum_{l,m}\left|\int \Psi_4^{lm}\mathrm{d}t\right|^2, \tag{B.50}$$

where the summation is taken over $l \geq 2$ and $-l \leq m \leq l$. This summation convention is used throughout this section. The computation of linear momentum flux involves the triple product including $l = 1, s = 0$ harmonics via s_i. The momentum fluxes in the x and y direction are conveniently expressed in terms of a complex variable $P_+ \equiv P_x + iP_y$ as

$$\frac{\mathrm{d}P_+}{\mathrm{d}t} = \lim_{r\to\infty}\frac{r^2}{8\pi}\sum_{l,m}\int \Psi_4^{lm}\mathrm{d}t$$
$$\times\int\left(a_{lm}\bar{\Psi}_4^{l,m+1} + b_{l,-m}\bar{\Psi}_4^{l-1,m+1} - b_{l+1,m+1}\bar{\Psi}_4^{l+1,m+1}\right)\mathrm{d}t, \tag{B.51}$$
$$\frac{\mathrm{d}P_z}{\mathrm{d}t} = \lim_{r\to\infty}\frac{r^2}{16\pi}\sum_{l,m}\int \Psi_4^{lm}\mathrm{d}t$$
$$\times\int\left(c_{lm}\bar{\Psi}_4^{lm} + d_{lm}\bar{\Psi}_4^{l-1,m} + d_{l+1,m}\bar{\Psi}_4^{l+1,m}\right)\mathrm{d}t, \tag{B.52}$$

where coefficients are given by

$$a_{lm} = \frac{\sqrt{(l-m)(l+m+1)}}{l(l+1)}, \tag{B.53}$$

$$b_{lm} = \frac{1}{2l}\sqrt{\frac{(l+2)(l-2)(l+m)(l+m-1)}{(2l+1)(2l-1)}}, \tag{B.54}$$

$$c_{lm} = \frac{2m}{l(l+1)}, \tag{B.55}$$

$$d_{lm} = \frac{1}{l}\sqrt{\frac{(l+2)(l-2)(l+m)(l-m)}{(2l+1)(2l-1)}}. \tag{B.56}$$

The angular momentum flux is obtained applying the angular momentum operator on the spin-weighted spherical harmonics, and given by

$$\frac{\mathrm{d}J_x}{\mathrm{d}t} = \lim_{r\to\infty} \frac{r^2}{32\pi}\mathrm{Im}\left[\sum_{l,m}\iint \Psi_4^{lm}\mathrm{d}t\mathrm{d}t' \right.$$
$$\left.\int \left(f_{lm}\bar{\Psi}_4^{l,m+1} + f_{l,-m}\bar{\Psi}_4^{l,m-1}\right)\mathrm{d}t\right], \tag{B.57}$$

$$\frac{\mathrm{d}J_y}{\mathrm{d}t} = -\lim_{r\to\infty} \frac{r^2}{32\pi}\mathrm{Re}\left[\sum_{l,m}\iint \Psi_4^{lm}\mathrm{d}t\mathrm{d}t' \right.$$
$$\left.\int \left(f_{lm}\bar{\Psi}_4^{l,m+1} - f_{l,-m}\bar{\Psi}_4^{l,m-1}\right)\mathrm{d}t\right], \tag{B.58}$$

$$\frac{\mathrm{d}J_z}{\mathrm{d}t} = \lim_{r\to\infty} \frac{r^2}{16\pi}\mathrm{Im}\left[\sum_{lm} m\iint \Psi_4^{lm}\mathrm{d}t\mathrm{d}t' \int \bar{\Psi}_4^{lm}\mathrm{d}t\right], \tag{B.59}$$

where $f_{lm} = \sqrt{(l-m)(l+m+1)}$.

Appendix C
Locating the Apparent Horizon and Computing an Approximate Killing Vector in the Initial Value Problem

To compute the mass and spin angular momentum of the BH in the initial data, especially those devoid of the axisymmetry, it is necessary to locate the apparent horizon (AH) and the approximate Killing vector (AKV) on the AH. In particular, the location of the AH is not known a priori in the puncture framework, whereas the location is prescribed by the boundary condition in the excision framework. In this appendix, we review the method to locate the AH and to compute an AKV on the AH in the initial value problem. In this appendix, the AH is denoted by $\mathscr{S}$.

C.1 The Apparent Horizon Finder

The location of the AH, which is defined as a two surface where the expansion of the outgoing null vector vanishes,[9] is determined using the method developed in [17]. The outgoing null vector is chosen to be

$$l^{\mu} = \frac{1}{\sqrt{2}}(n^{\mu} + s^{\mu}), \tag{C.1}$$

where s^{μ} is the spatial, outward unit normal vector to the AH, and the expansion of l^{μ} is defined by

$$\begin{aligned} \boldsymbol{\Theta}_{(l)} &\equiv q^{\mu\nu}\nabla_{\mu}l_{\nu} \\ &= \frac{1}{\sqrt{2}}(D_i s^i - K + K_{ij}s^i s^j). \end{aligned} \tag{C.2}$$

Hence, we have to solve the equation

[9] Rigorously saying, this is the definition of the marginally outer trapped surface. For the marginally outer trapped surface to be an AH, the expansion of the ingoing null vector have to be non-positive.

K. Kyutoku, *The Black Hole–Neutron Star Binary Merger in Full General Relativity*, Springer Theses, DOI: 10.1007/978-4-431-54201-8,

$$D_i s^i - K + K_{ij} s^i s^j = 0 \tag{C.3}$$

in order to find the location $r = h(\theta, \varphi)$ of the AH. Here, it is already assumed that the AH is topologically S^2, and is not distorted too much to be represented by a single-valued function of angular coordinates [18]. For later convenience, it is useful to rewrite this equation as

$$q^{ij}(D_i s_j - K_{ij}) = 0, \tag{C.4}$$

where $q_{ij} \equiv \gamma_{ij} - s_i s_j$, which is the induced metric on $\mathscr{S}$. It is also useful to realize that the AH is considered to be a constant surface of the function,

$$F(r, \theta, \varphi) \equiv r - h(\theta, \varphi). \tag{C.5}$$

Because the unit normal s_i is defined as

$$s_i \equiv \frac{D_i F}{\sqrt{\gamma^{jk}(D_j F)(D_k F)}}, \tag{C.6}$$

the exact unit normal to the AH is unknown until the location of the AH is specified. Inversely, this equation is solved to find a correct function $h(\theta, \varphi)$ by an iterative method, starting from some initial guess for $h(\theta, \varphi)$. Because the initial value problem itself is solved by the iterative method, the initial guess is chosen to be the function at the previous step with the exception for the very initial step, at which we choose $h = M/2$. Equation (C.4) becomes

$$q^{ij}\left[\frac{D_i D_j F}{\sqrt{\gamma^{kl}(D_k F)(D_l F)}} - K_{ij}\right] = 0, \tag{C.7}$$

using $q^{ij} s_j = 0$. Defining the difference between D_i and $\overset{\circ}{D}_i$ as

$$\Delta^k{}_{ij} \equiv \frac{1}{2}\gamma^{kl}\left(\overset{\circ}{D}_i \gamma_{jl} + \overset{\circ}{D}_j \gamma_{jl} - \overset{\circ}{D}_l \gamma_{ij}\right), \tag{C.8}$$

the conformal transformation of this equation leads

$$\frac{\overset{\circ}{D}{}^i \overset{\circ}{D}_i F}{\psi^4 \sqrt{\gamma^{kl}(D_k F)(D_l F)}} - \frac{s^i s^j \overset{\circ}{D}_i \overset{\circ}{D}_j F}{\sqrt{\gamma^{kl}(D_k F)(D_l F)}} - q^{ij}\left(\frac{\Delta^m{}_{ij} \overset{\circ}{D}_m F}{\sqrt{\gamma^{kl}(D_k F)(D_l F)}} + K_{ij}\right) = 0. \tag{C.9}$$

It can be shown by a little algebra that $\overset{\circ}{D}{}^i \overset{\circ}{D}_i F$ is expanded as

$$\overset{\circ}{D}{}^i\overset{\circ}{D}_i F = -\frac{1}{\psi^4\sqrt{\gamma^{ij}(D_iF)(D_jF)h^2}}(\Delta_{\theta\varphi} - 2)h, \tag{C.10}$$

where $\Delta_{\theta\varphi}$ is the angular Laplacian operator given by

$$\Delta_{\theta\varphi} \equiv \frac{\partial^2}{\partial\theta^2} + \frac{\cos\theta}{\sin\theta}\frac{\partial}{\partial\theta} + \frac{1}{\sin^2\theta}\frac{\partial^2}{\partial\varphi^2}. \tag{C.11}$$

Finally, we obtain an elliptic-type equation to determine $h(\theta, \varphi)$ as

$$(\Delta_{\theta\varphi} - 2)h = -\psi^4 h^2\sqrt{\gamma^{kl}(D_kF)(D_lF)} \times \left\{\frac{s^i s^j \overset{\circ}{D}_i\overset{\circ}{D}_j F}{\sqrt{\gamma^{kl}(D_kF)(D_lF)}} + q^{ij}\left(\frac{\Delta^m{}_{ij}\overset{\circ}{D}_m F}{\sqrt{\gamma^{kl}(D_kF)(D_lF)}} + K_{ij}\right)\right\}. \tag{C.12}$$

For example, this equation is actually satisfied for the Schwarzschild BH in the isotropic coordinates,

$$h(\theta, \varphi) = \frac{M}{2}, \ \psi = 1 + \frac{M}{2r}, K_{ij} = 0. \tag{C.13}$$

C.2 The Approximate Rotational Killing Vector

Because the BH–NS binary spacetime is not stationary nor axisymmetric, any Killing vector does not exist. Especially, there is no rotational Killing vector around the BH. However, the BH is believed to be in an approximately axisymmetric state intrinsically, if the distance between the BH and NS is fairly large and tidal distortion is not severe. In this situation, it may be possible to define an approximate rotational Killing vector on the BH horizon in a quasilocal manner. For this purpose, we adopt the method developed in [19] to compute an AKV with the normalization condition proposed in [20]. There is another commonly-used method proposed in [21], but the method adopted in this study is advantageous in that the obtained AKV satisfies the divergence-free condition by construction (see below).

Because we are trying to find a vector field which is close to the Killing vector field, we want to have a vector field satisfying at least some part of the Killing equation,

$$\mathscr{D}_A\phi_B + \mathscr{D}_B\phi_A = 0. \tag{C.14}$$

It is known that, to define a gauge-invariant angular momentum in the isolated horizon framework, the approximate Killing vector must be divergence free [22], and the divergence-free property results from the trace of the Killing equation. Therefore, we assume ϕ^A to be expressed in a general form of the divergence-free vector field on a two surface,

$$\phi^A = \epsilon^{AB}\mathscr{D}_B v, \tag{C.15}$$

where v is a scalar function on $\mathscr{S}$, and try to minimize the norm of the shear defined as

$$\|\sigma\|^2 \equiv \oint_{\mathscr{S}} \sigma_{AB}\sigma^{AB}\sqrt{q}\mathrm{d}^2x, \sigma_{AB} \equiv \frac{1}{2}(\mathscr{D}_A\phi_B + \mathscr{D}_B\phi_A - q_{AB}\mathscr{D}_C\phi^C), \tag{C.16}$$

by a variational method to find a vector field which is the closest to the Killing vector field. Minimizing $\|\sigma\|^2$ is equivalent to minimizing the residual, trace-free part of the Killing equation. If we want to solve this variational problem with respect to v, however, we may only obtain $v = \text{const}$ and $\phi^A = 0$, which is a trivial kernel of the variational operator. Therefore, we adopt the Lagrange-multiplier technique introducing the multiplier λ and a constant N, so that the norm of the approximate Killing vector is fixed to a nonzero value. Specifically, we require the normalization condition,

$$\oint_{\mathscr{S}} \mathscr{R}\phi_A\phi^A\sqrt{q}\mathrm{d}^2x = N, \tag{C.17}$$

where the weight $\mathscr{R}$ is chosen to make the integrand nondimensional, to be satisfied. Integrating by parts, the functional we want to minimize becomes

$$\begin{aligned}&\oint_{\mathscr{S}} v\left[\mathscr{D}^4 + \mathscr{R}\mathscr{D}^2 + (\mathcal{D}^A\mathscr{R})\mathscr{D}_A\right]v\sqrt{q}\mathrm{d}^2x\\&\quad -\lambda\left\{\oint_{\mathscr{S}} v[\mathscr{R}\mathscr{D}^2 + (\mathscr{D}^A\mathscr{R})\mathscr{D}_A]v\sqrt{q}d^2x + N\right\},\end{aligned} \tag{C.18}$$

and the variation with respect to v gives

$$[\mathscr{D}^4 + \mathscr{R}\mathscr{D}^2 + (\mathscr{D}^A\mathscr{R})\mathscr{D}_A]v = \lambda[\mathscr{R}\mathscr{D}^2 + (\mathscr{D}^A\mathscr{R})\mathscr{D}_A]v. \tag{C.19}$$

The Lagrange multiplier, λ, is determined to be the smallest eigenvalue of this variational problem. In the actual computation, we solve this problem by rewriting as

$$\begin{aligned}&\mathscr{D}^2L = (1-\lambda)\left[\frac{1}{2}\left(\mathscr{D}^A\mathscr{R}\right)(\mathscr{D}_A v) - \mathscr{R}L\right],\\&\mathscr{D}^2 v = -2L,\end{aligned} \tag{C.20}$$

where L is related to the derivative of the Killing vector by

$$\mathcal{D}_A\phi_B = L\varepsilon_{AB} + \sigma_{AB}. \tag{C.21}$$

Notice that the value of N is not determined in this procedure. For example, the exact solution for a flat two surface with a constant radius r is given in terms of the spherical harmonics Y^{10} by

$$v = r^2 \cos\theta, L = \cos\theta, \tag{C.22}$$

where $\mathscr{R} = 2/r^2$, up to a constant scaling. In practice, we decompose the induced metric into flat and deviation parts as

$$\sqrt{q}q^{AB} = f^{AB} + r^{AB}, \tag{C.23}$$

and rewrite $\mathscr{D}^2$ so that the operator becomes the sum of an angular Laplacian and correction terms.

We next have to fix the normalization of v, of which the scaling was unconstrained up to a constant. For this purpose, it is useful to introduce a coordinate system (v, φ) on $\mathscr{S}$. If the approximate Killing vector is parametrized by a parameter τ as $\phi^A = (\mathrm{d}/\mathrm{d}\tau)^A$, for which the integral curve $\mathscr{C}$ of ϕ^i corresponds to the contour line of v by definition, it can be shown that the increase of τ along one circle of $\mathscr{C}$ parametrized by v is

$$\begin{aligned} \tau(v) &= \oint_{\mathscr{C}} \frac{\mathrm{d}\varphi}{\phi^{\varphi}(v,\varphi)} \\ &= \oint_{\mathscr{C}} \sqrt{q} d\varphi, \end{aligned} \tag{C.24}$$

and therefore averaging this on $\mathscr{S}$ yields

$$\langle\tau\rangle = \frac{A_{\mathrm{AH}}}{v_{\max} - v_{\min}}. \tag{C.25}$$

We fix the normalization of v by requiring $\langle\tau\rangle = 2\pi$. In practice, we require

$$\oint_{\mathscr{S}} (v - \langle v\rangle)^2 \sqrt{q} \mathrm{d}^2 x = \frac{A_{\mathrm{AH}}^3}{48\pi^2}, \tag{C.26}$$

which holds for the Kerr BH (including a coordinate sphere in the flat spacetime), for the sake of a numerical accuracy.

References

1. R. Arnowitt, S. Deser, C.W. Misner, in *Gravitation*, ed by L. Witten (Wiley, New York, 1962), pp. 227–265
2. R.M. Wald, *General Relativity* (The university of Chicago press, 1984)

3. E. Poisson, *A Relativist's Toolkit: The Mathematics of Black-Hole Mechanics* (Cambridge university press, 2004)
4. J.W. York, Phys. Rev. Lett. **28**, 1082 (1972)
5. G.W. Gibbons, S.W. Hawking, Phys. Rev. D **15**, 2752 (1977)
6. J.W. York, in *Gravitational Radiation*, ed by L. Smarr (Cambridge University Press, Cambridge, 1979), p. 83
7. J.D. Brown, J.W. York, Phys. Rev. D **47**, 1407 (1993)
8. T. Regge, C. Teitelboim, Ann. Phys. (NY) **88**, 286 (1974)
9. J. Katz, D. Lynden-Bell, W. Israel, Class. Quantum Gravity **5**, 971 (1988)
10. S.W. Hawking, G.T. Horowitz, Class. Quantum Gravity **13**, 1487 (1996)
11. M. Maggiore, *Gravitational Waves Volume 1: Theory and Experiments* (Oxford university press, 2008)
12. M. Alcubierre, *Introduction to 3+1 Numerical Relativity* (Oxford science publications, Oxford, 2008)
13. R.A. Isaacson, Phys. Rev. **166**, 1263 (1968)
14. R.A. Isaacson, Phys. Rev. **166**, 1272 (1968)
15. D.R. Brill, J.B. Hartle, Phys. Rev. **135**, B271 (1964)
16. K.S. Thorne, Rev. Mod. Phys. **52**, 299 (1980)
17. L.M. Lin, J. Novak, Class. Quantum Gravity **24**, 2665 (2007)
18. S. Bonazzola, E. Gourgoulhon, J.A. Marck, Phys. Rev. D **58**, 104020 (1998)
19. G.B. Cook, B.F. Whiting, Phys. Rev. D **76**, 041501 (2007)
20. G. Lovelace, R. Owen, H.P. Pfeiffer, T. Chu, Phys. Rev. D **78**, 084017 (2008)
21. O. Dreyer, B. Krishnan, D. Shoemaker, E. Schnetter, Phys. Rev. D **67**, 024018 (2003)
22. A. Ashtekar, J. Engle, T. Pawlowski, C. van den Broeck, Class. Quantum Gravity **21**, 2549 (2004)

MIX
Papier aus verantwortungsvollen Quellen
Paper from responsible sources
FSC® C105338

If you have any concerns about our products,
you can contact us on
ProductSafety@springernature.com

In case Publisher is established outside the EU,
the EU authorized representative is:
Springer Nature Customer Service Center GmbH
Europaplatz 3, 69115 Heidelberg, Germany

Printed by Libri Plureos GmbH
in Hamburg, Germany